TEAM TOYOTA

SUNY Series in the Sociology of Work
Richard Hall, editor

TEAM TOYOTA

Transplanting the Toyota Culture
to the Camry Plant in Kentucky

Terry L. Besser

State University of
New York Press

Quoted material on pages 85 and 86 was taken from *Working for the Japanese: Inside Mazda's American Auto Plant* by Joseph T. Fucini and Suzy Fucini. Copyright © 1990 by Joseph J. Fucini and Suzy Fucini. Reprinted with permission of The Free Press, a division of Simon & Schuster, Inc.

Published by
State University of New York Press

© 1996 State University of New York

Printed in the United States of America

For information, address the State University of New York Press,
State University Plaza, Albany, N.Y., 12246

Production by Bernadine Dawes • Marketing by Theresa Abad Swierzowski

Library of Congress Cataloging-in-Publication Data

Besser, Terry L., 1945–
      Team Toyota : transplanting the Toyota culture to the Camry plant
in Kentucky / Terry L. Besser.
            p.   cm. — (SUNY series in the sociology of work)
      Includes bibliographical references and index.
      ISBN 0–7914–3145–2 (hc : alk. paper). — ISBN 0–7914–3146–0 (pb :
alk. paper)
      1. Automobile industry workers—Kentucky—Georgetown.   2. Toyota
Jidōsha Kōgyō Kabushiki Kaisha—Employees.   3. Industrial sociology—
Kentucky—Georgetown—Case studies.   4. Automobile industry and
trade—Kentucky—Georgetown—Management—Case studies.
5. Corporations, Japanese—Kentucky—Georgetown—Management—Case
studies.   I. Title.   II. Series.
HD6956.A82U63   1996
331.7'6292'09769425—dc20
                                                                96–11692
                                                                CIP

10 9 8 7 6 5 4 3 2

Contents

Foreword / vii

Chapter 1. Introduction / 1

Chapter 2. Method of Study / 27

Chapter 3. The Context / 35

Chapter 4. Team Toyota / 49

Chapter 5. On Restriction / 119

Chapter 6. On Automatic / 139

Chapter 7. The Women of Toyota / 151

Chapter 8. Conclusion / 173

Appendix A / 185

Appendix B / 187

References / 189

Index / 197

Foreword

Last year I was considering buying a Toyota. Trying to be an informed consumer, I called my niece, who is an engineer at one of the major U.S. auto firms. She said that her company watches Toyota products very carefully, since they consistently make high-quality cars and trucks.

Here, in *Team Toyota,* we have another high-quality product. Besser combines a strong theoretical basis with insightful data from interviews and observations. All of this is presented in a most engaging manner—all too rare in sociological writing.

Two basic issues are addressed. First: What happens when the Japanese form of organization and management style is exported to the United States? The answer is complex. Some things are changed, while others remain very Japanese. In her analysis, Besser widens her scope and notes that the Japanese approach has not been exported in all industries in which Japanese firms have established branch ventures in the United States. This leads me to the conclusion that there is no one best way to organize or manage. Single and simple prescriptions just do not work.

The second issue is equally complex: What happens to U.S. workers who choose to work in this setting? This question receives the most attention in this book. The workers speak in their own words throughout the book and they speak eloquently.

The workers do not speak in a single voice. Some are very happy with the job security they believe they have. Others have become injured or ill from their work. Some enjoy the team concept as it engulfs their work life. Others find the teams overwhelming.

Besser adds her own voice to those she has recorded and transcribed. The final product is a careful analysis and interpretation of the Camry plant in Kentucky. Besser neither uniformly criticizes nor constantly praises the Toyota operation. Instead, she carefully interprets her findings with insights gained from her observations and interviews in the plant. Her interpretations are informed by the growing literature on modern Japanese organizations. They are also informed by Besser's own keen insights.

Besser adds an issue that has been missing in earlier studies of Japanese transplant organizations. Her chapter on gender issues in the Toyota plant is very informative and balanced in that she reports on both U.S. and Japanese women as they work in and are influenced by the transplant organizational form.

Richard H. Hall

1

INTRODUCTION

Toyota is arguably the quintessential Japanese organization. It comes as close as any firm does to practicing "Japanese management" that includes features such as lifetime employment, company ideology, and group responsibility. Despite its size and power, it hangs on to the rural, family-owned image. Technically, this image is certainly true. The Toyoda family still directs and owns most of the corporation, and Toyota's Japanese manufacturing facilities are located in rural Aichi prefecture. However, in reality, Toyota is one of the world's corporate giants and bears little resemblance to the average rural family-owned business. It coordinates an extensive supplier system that acts to externalize many of the risks and costs of modern mass production, has staggering in-house financial resources, and utilizes aggressive marketing strategies. It has developed a production system that, according to Womack, Jones, and Roos (1990) of MIT and Kenney and Florida (1993), represents the first major innovation in the production of automobiles since Ford's assembly line innovations. Few Japanese organizations have been as successful as Toyota. In fact, few organizations in the world enjoy Toyota's prosperity and power.

Perhaps because of its success, Toyota has been unwilling to change the essentials of its production and management systems in its overseas operations. Like the other Japanese automobile companies who are all recent arrivals to the U.S. manufacturing scene, Toyota intends to transplant a modified version of its Japanese style of organizing to its U.S. plants. This is in sharp contrast to the Japanese trading companies, banks, and electronics manufacturers that have been in the United States since the 1960s and have used standard Western management practices in their U.S. facilities. Thus, the Japanese automobile companies' investment in the United States provides organizational scholars with the unparalleled opportunity to study the creation of archetypical Japanese organizations in a "foreign environment" and the intersection of Japanese management with American employees. The Japanese automobile plants in the United States are actually large organizational laboratories conducting social and organizational experiments. No wonder the transplants have generated such a significant amount of public and scholarly interest.

I was fortunate to be at the University of Kentucky in Lexington when Toyota established its Camry plant—Toyota Motor Manufacturing (TMM), located about fifteen miles north of Lexington in Georgetown. I was able to study Toyota and TMM during this critical time. The purpose of this book is to present the results of that research: an analysis of how Toyota is re-creating its form of management and team culture in its Georgetown plant.

I began by reading the material currently available on the topic of Japanese management. In this way I was able to define what is meant by "Japanese management." Next, I needed to find out what the employees were experiencing at TMM and discover the official policies, structures, procedures, hierarchies, and so forth, in the organization. With this information I could determine the degree of overlap between the "ideal of Japanese management" and the situation at TMM. More importantly though, this exercise allowed me to show how abstract organizational features like group responsibility, culture, communications patterns, and the like, become concrete daily activities of employees via the organizational mechanisms of job descriptions, personnel policies, departmental structure, hiring and training procedures, and so forth.

To organize and make sense of the data I gathered in my research, I developed a theoretical framework using "team" as the dominant metaphor. I was very reluctant to use the concept of team, as it is a theme heard too often in Toyota commercials—I feared that my theory would sound more like an advertisement than a research treatise. Nonetheless, it was impossible to avoid. "Team" permeates every aspect of the organization and is a large part of the daily reality of TMM workers. Team as metaphor provides a means of articulating between the levels of reality at TMM—that is, between the worker, the work team, the local company, and corporate Toyota. It illustrates the interrelationships and interdependencies between these levels. Using the team theory we are able to see how, in fact, Japanese organizations penetrate the small work group and recruit small group support for organizational goals. We realize what organizational mechanisms contribute to the creation of the "community-of-fate" culture among employees. We can predict areas of tension and potential problems. Furthermore, the theory can provide employees with the conceptual tools to better understand the organization in which they are embedded.

Bureaucracy: Traditional Western Management

Before examining the characteristics of Japanese management, let us first consider the standard model of management used in the West. In that way it will be clear just how Japanese management differs from common Western practice. The best and most often cited depiction of this model was presented by Max Weber in the 1920s in his elaboration of the characteristics of bureaucracy. Weber made it clear that he was describing an ideal type; that is, he was presenting the features of the perfect bureaucracy. He recognized that this model probably did not (and still does not) exist anywhere in reality. However, it is still important to understand it since, no matter how imperfectly, people and organizations continuously aspire to implement the model.

Most people view bureaucracy as the least ideal way to organize complex human activity. Everyone loves to hate bureaucracy. However, its negative image results from misconceptions. There are two reasons for the general disdain of bureaucracy. The first is that most people don't know the real meaning of bureaucracy and believe it is synonymous with government operations (which, according to common opinion are ipso facto bumbling and ineffective). Second, bureaucracy as an organizing format is so successful that almost all organizations use it and, therefore, we don't realize how bad things would be without it. Bureaucracies are ubiquitous in modern life and most admonitions for organizational improvement are really calls for more bureaucracy, not less.

One final point before examining Weber's model. Charles Perrow suggests that organizations are tools, and it is very helpful to use this metaphor in our examination of bureaucracy. Accordingly, bureaucracies are tools used by masters to accomplish some goal. The masters decide what the goal should be, but are not themselves part of the bureaucracy. The board of directors of a corporation, the executive board of the Girl Scouts or the United Way, the local school board, the owner of a professional football team, and the legislature are all masters who determine policy. The bureaucracies they control are charged with the mission of carrying out these policies in an efficient manner.

Let us turn now to the basic characteristics of Weber's ideal type bureaucracy. Perhaps the central feature of bureaucracy, as Weber presented it, is its pyramidal structure. At the top of the pyramid is the chief administrator (sometimes called the general manager, hospital administrator, superintendent, secretary of defense, chief executive officer, etc.) who is appointed by the policy-setting body to accomplish their goals. At the bottom are the functionaries (teachers, assembly workers, doctors, soldiers, ministers, accountants, waitresses, etc.) who actually perform the work of the organization. In between are lower-level administrators, vice presidents, principals, shop-floor managers, sergeants, bishops, and so forth, who act as conduits passing orders from the chief administrator to the functionaries, and communication and information back up from the functionaries to the chief administrator. These intermediary positions use their expertise to translate general goals into the specific concrete actions needed to accomplish the goals. With each step down the pyramid the general goal becomes more specific, more refined. For example, what started out as a general policy goal of making 4 percent profit is divided into more specific departmental goals like more aggressive marketing, training employees, buying new machinery, or closing less profitable outlets. Ultimately, these goals are transformed into policies, rules, and procedures that are enacted in the day-to-day activities of the functionaries. Some positions in a bureaucracy are supportive or advisory and are not inside the pyramid. Researchers, legal advisors, and clerical positions are examples of what are called "staff positions," as opposed to the line positions contained within the pyramid. The most important distinction between line and staff positions is that, theoretically, people in line positions can be promoted up through the pyramidal hierarchy. People in staff positions usually cannot be.

Lines of authority and responsibility are clearly specified in a bureaucracy. There is a chain of command. Everyone in the pyramid knows they report to the position directly above them and the division of labor ensures that each identified task has one and only one person responsible for it. Therefore, credit and blame can be doled out accordingly. Those of us who have worked in a nonbureaucratic organization like a mom and pop restaurant or grocery store know how difficult it is to do your job when all members of the owning family are in charge. It's not clear who you report to and, in fact, situations occur when two or three of the bosses give conflicting orders. Similarly, if servers in a restaurant do not know which tables to wait on, conflict and ambiguity can result, affecting the quality of service to customers.

In a bureaucracy people are hired for their expertise and promoted for meritorious performance. This requires that each organizational task be outlined in job descriptions that contain the exact responsibilities of the job and what quantifiable measure of expertise and ability apply. Prior work experience, paper and pencil tests, and academic credentials often serve as indicators of expertise. Since each position has clearly identified responsibilities, how an individual fulfills those responsibilities is a measure of merit and provides the basis for promotion decisions. This principle is intended to lessen the subjectivity that sometimes affects hiring and promotion decisions so that people with the ability and knowledge to accomplish the organization's mission (not cousins, friends, or people who helped the chief administrator get into office) will be hired for organizational jobs; and office holders will be rewarded with promotion, or salary increases, for exemplary performance in fulfilling the responsibilities of their job (not for how loyal they are to the chief administrator). The reasoning is that if people are hired for their skill in performing a job—teaching, for example—the mission of the organization (in this example, educating students) is more likely to be accomplished.

Another characteristic of bureaucracy is that administration is a full-time job and constitutes a special skill. That administration is a separate skill is affirmed by the creation of whole bodies of knowledge and corresponding academic disciplines—for example, management and public administration. The purpose of these programs is to prepare their graduates so that they can administer a broad range of organizations, irrespective of those organizations' specific goals.

In a bureaucracy the functionaries do not administer while performing their primary tasks. And although administrators may have worked their way up from the rank of functionary, they do not now routinely engage in that kind of work. However, the more emphasis an organization places on the science of administration (as opposed to the profession of the functionaries), the less promotion from the ranks is likely to occur. In many for-profit organizations it is not necessary to start on the line, or in the ditches, or behind the counter to move up the management ranks. In fact, administrative skill is often viewed as completely distinct from the skills necessary to carry out the mission of the organization. Many believe that a good administrator in business can administer a government agency, even though that person has no experience performing the agency's work, or that serving as an officer in the military qualifies a person to be a manager in business, and so on.

There are, however, limitations to the transferability of management personnel and skills between organizations. In organizations where the mission of the organization is more important than, and may conflict with, efficient administration, the skills of the functionaries are viewed as more important than administrative skill. Hospitals, scientific research institutes, and schools are examples of organizations in which there could be conflicts between efficiency and the organization's mission. Thus, many people feel ambivalent about professional managers running hospitals (Could an army general effectively manage a hospital?), or noneducators directing school districts (Would the CEO of General Motors be able to effectively administer the Detroit school system?), and recognize the absurdity of professional administrators overseeing a religious order. In these circumstances, special training is given to selected functionaries creating linking professions. Hospital administration, educational administration, and public administration are college curriculums (and professions) intended to combine the skill and understanding of the functionaries with the knowledge of the science of administration.

People who work in a bureaucracy view their job as a vocation; that is, they are dedicated to the mission of the organization. Their loyalty is to the organization (and its goals), not to any particular chief administrator. To ensure that this occurs, office holders must have some kind of job security. They must be protected from capricious or malicious retribution from chief administrators. Job security allows office holders to challenge chief administrators for the sake of the organizational mission. It is intended to enable office holders to see themselves as servants of the organizational goals and not as personal servants of the chief administrator.

On the other hand, the chief administrator usually enjoys no such job security. The policy-making body (the master) reserves for itself the right to summarily replace chief administrators, who are ultimately held responsible for the success or failure of organizational goal achievement. Congress and the president can remove the secretary of defense, but individual soldiers, even admirals and generals, are protected from political removal. The board of regents of a university can remove the president of the university, but not particular faculty members. This is not to say that these office holders cannot be removed. They can be, but the procedure is lengthy, requires proof of malfeasance of office (court martial, disbarment, censure, etc.), and often must be approved by the office holder's peers. This characteristic occurs less frequently in private, for-profit bureaucracies than in public ones. Even so, white-collar employees of large corporations enjoyed a kind of "permanent employment" until the last decade, and union representation prevented capricious or malicious dismissal of blue-collar employees.

Another principle of bureaucracy is that office holders are entitled to use the resources (including the power) of the organization only for organization business and only to achieve organizational goals. Prior to bureaucracy becoming commonplace, people often used church, government, and business organizational resources for their own personal ends, to achieve personal goals. In businesses there was no separation of monies spent for tools, material, and labor from those spent for personal and family reasons. Neither was there a well-defined limit to what organizations could ask of office holders. Office holders' families sometimes were held accountable for the office holder's mistakes and might have to fulfill the duties of the job should the office holder become sick or injured. Family and personal financial resources could be levied for organizational activities.

The difference is illustrated by considering a family-owned business that is not bureaucratized and not managed by hired professionals. Members of the family work in the business whenever there is a need. They use whatever they want from the company stores. There really is no separation between personal and organizational business. Under these circumstances, it is very difficult to tell how much profit the business earns, where it could be made more efficient, and where capital should be invested because uncounted hours of labor are contributed by family members and unknown amounts of business resources are used for personal ends.

Probably the principles that most people readily associate with bureaucracy are that (1) they are managed by strict adherence to written rules and standard operating procedures and (2) they make use of files in the performance of their tasks. Horror stories abound about how bureaucracies turn individual clients into numbered files, dehumanize them, and then misfile or lose the file. Similarly, the absurdity and inflexibility that can result from a preponderance of rules, and strict adherence to those rules, is legendary. However, consider for a moment the alternatives. Written records of transactions and files immensely improve the memory of organizations, allowing them to conduct business in a consistent manner beyond the life of any one office holder. This becomes an important factor when you need to prove to someone that you graduated from college, pay your bills, were married to a particular person, or that you were born on a certain date, to specific parents, and were given the name you've grown accustomed to using.

Before European governments were bureaucratized, capitalists were already using bureaucratic procedures and lobbying for governments to follow suit. The reason capitalists favored the bureaucratic form was that adherence to rules and recorded transactions would help to ensure that government activities would be predictable and even-handed, and that these features would exist no matter who held the offices. At the same time, people began to believe that "everyone should be equal before the law." Therefore, they wanted government to treat all citizens, regardless of wealth or family (and today regardless of sex, race, ethnicity, or religion), the same. What better way to do that than to turn all clients into numbers and to enforce the rules without regard to personal situation?

Bureaucracy is an extremely popular method of organizing because, as Weber put it, "It is a power instrument of the first order." There is no other method of organizing complex human activities that is as efficient at implementing the master's orders—and as enduring and persistent—as a well-functioning bureaucracy. All the popular lore about the ineptitude of bureaucracy is usually about the imperfect articulation of bureaucracy in an organization. To illustrate the point that bureaucratic and bungling are not synonymous, let us consider one of the most effective bureaucracies around today, professional baseball teams. Satisfy yourself that professional baseball teams are indeed bureaucracies. They are organized pyramidal with functionaries (baseball players) on the bottom and full-time professional managers above, decreasing in number until the general manager. There is a strict division of labor, clearly specified responsibilities, decisions go from the top of the pyramid to the bottom, positions are filled on the basis of merit, and bonuses or raises are accorded to those who do the best job of fulfilling their job responsibilities. Professional baseball has elaborate rules and regulations and maintains extensive records on games and players down to the most minute details of batting and pitching performances.

Two ingredients of bureaucracies are missing from professional baseball, job security and a career ladder. Movement from player to staff or team management occurs for only a minority of players and is not always viewed as a promotion. Nor is it always accompanied by an increase in pay. Some might contend that job security is the critical feature of bureaucracy and that an organization without job security is not a bureaucracy. I note that position. However, the advantage of using professional sports as exemplars of bureaucracy is that their organizational goals are much more unambiguous and easy to measure than in most other types of organizations. Games are won or lost. Seasons are winning or losing. This advantage outweighs the fact that professional sports teams may be an imperfect match with the ideal type bureaucracy. Recall that no organization perfectly replicates all the characteristics of bureaucracy.

In case there are doubts about the superior ability of a bureaucratic baseball team to achieve the master's goal, which usually is to win baseball games, imagine a contest between the Cincinnati Reds and a nonbureaucratic baseball team, like the neighborhood crew who get together on some Sunday afternoons to play baseball. Who would win? Of course, you might think the Reds perform better not because of the way they are organized but because of the huge salaries given to players. How then to account for the fact that the local college or high school baseball team would almost certainly defeat the nonbureaucratic neighborhood team also? Unquestionably, part of the difference has to do with the talent of the players. An essential feature of bureaucracy is filling positions with people who are the most skilled in the requirements of the job. However, the success of a bureaucratic baseball team results from more than recruiting the best talent. If the most gifted athletes in the world are hired but they rotate positions haphazardly, are expected to raise money and make travel arrangements, do not know or adhere to the rules, have no one to organize practices or strategies, are rewarded capriciously partially because no one is keeping track (recording and filing) of performance, and there is constant bickering since no one wants to catch and everyone wants to pitch and all want to bat lead off (there is no one to make decisions), how effective will they be at winning baseball games? The most convincing argument that successful baseball teams are bureaucracies is evident in the answer to the question, What would have to be done to whip the neighborhood team into shape to compete effectively against the Reds? The answer, bureaucratize.

Although I have done my best to counter the negative stereotype of bureaucracy that seems to prevail in the common wisdom of our culture, I would be remiss if I left the reader with the impression that bureaucracy is without weaknesses. It is not my purpose here to delineate all the advantages and disadvantages of bureaucracy. Sufficient to say that there are weaknesses, the more serious of which are not necessarily the ones that come readily to many people's minds. Effective bureaucracies must be constantly vigilant and even then sometimes the multiplication of rules leads to permanent paralysis. Worse still, bureaucracies tend to become so powerful that the master can no longer control them. Their relative immortality and expertise makes it difficult for masters to effect any fundamental internal changes. This can happen in for-profit bureaucracies, however market forces will usually provide an eventual correction. Public bureaucracies and private, nonprofit bureaucracies have no similar built-in destruct mechanism. The Chinese bureaucracy kept functioning as it always had in spite of the conquest of the empire by the Mongols. The Roman bureaucracy continued to collect taxes and distribute food and other supplies long after the central empire itself had collapsed. The bureaucracy of the Roman Catholic Church has continued to operate in pretty much the same way for almost two thousand years, surviving scandals, wars, plagues, and tumultuous disputes over church leadership and sacred doctrines.

Regardless of problems with bureaucracies, I contend that it is the prevailing standard method of organizing complex human activity in the West. With an understanding of the standard bureaucratic model, it will become clear in what way the Japanese style of management deviates.

Japanese Management

Japanese management refers to a particular kind of management characteristic of large Japanese organizations. As scholars of Japanese organizations like William Ouchi or Richard Pascale and Anthony Athos point out in their 1981 books, this style of management is not unique to Japanese firms. The practices ascribed to Japanese management are seen in a variety of non-Japanese organizations, but are more common and more normatively accepted in Japan. The number of organizations that implement almost all of the features of Japanese management is limited, even in Japan. Robert Cole (1979) and David Plath (1983), two researchers who have studied Japanese organizations in Japan, conclude that only about 30 percent of workers in Japan work for an organization that practices Japanese management. Thus, while variations of Japanese management are present in most Japanese organizations familiar to Americans and may be viewed by the Japanese as the model or ideal management style, in actuality the majority of Japanese workers do not experience Japanese management.

James Abegglen (1958), Michael Yoshino (1968), and Ronald Dore (1973) are the scholars whose work is usually recognized as providing the definition of Japanese management. They were studying Japanese organizations before it became a popular topic. My account of the ideal type Japanese organization is an aggregation of their descriptions. I will proceed by comparing Japanese management with the bureaucratic model—first by considering how the two differ, then I will show how they are similar, and finally, I will examine aspects of Japanese management that have no counterpart in the bureaucratic model.

Pyramidal Structure and Nenko
(The Seniority Pay and Promotion System)

Bureaucracies and Japanese organizations are both structured pyramidal, but the shape of the pyramid and how internal processes are associated with the pyramid differ in each. Japanese organizational pyramids are "tall, thin, and finely grained." Traditional bureaucratic pyramids are flatter and fatter at the bottom. The feature that is probably most directly responsible for the tall Japanese pyramid is the fact that promotion and pay are based on seniority and not merit. Recently many

Japanese firms have begun to consider merit as an element in their salary determination formula, and the significance of seniority in promotion and pay considerations declines at upper management levels where merit plays a larger role. Nonetheless, seniority remains important and has historically played a critical role in making Japanese organizations unique.

Since seniority is the central element in promotion, each group of employees hired at the same time is promoted at the same point in their career, and each year of seniority is rewarded with a slightly higher position and/or a larger salary than individuals hired in subsequent years. In order for the organization to accommodate higher status and pay for each level of seniority, it must have many positions to which people can be promoted. The levels do not have to differ from each other greatly to achieve the desired effect. The result is an organization with many levels in the hierarchy (finely grained) and relatively few people at each level, causing the organization chart to look taller and thinner than hierarchies in Western bureaucracies.

Consensus Decision Making

In bureaucratic pyramids, decisions are made at the top and are communicated to the lower levels where they are carried out. Feedback from workers and information about consumers, government regulation, competition, and so forth, moves from the bottom to the top where decisions can be made informed by this input from below. In contrast, in Japanese organizations, decisions are made by groups of people close to the issue under consideration. The process is supposed to work like this: anyone with an idea or concern can assemble the relevant people. These people meet, discuss the matter, get more information if necessary, deliberate, and all agree on whatever decision is reached. After the decision is arrived at, notice of the decision is circulated to other departments and climbs the hierarchical ladder to "inform" the higher levels. Higher management will usually rubber stamp approval unless the decision conflicts with decisions or matters in other parts of the organization of which the group is unaware. The decision then returns to the group for implementation.

Proponents of Japanese management concede that this process of decision making requires more time to reach decisions. However, because implementation is smoother and faster than in Western bureaucracies, they contend that on balance no more time is expended in the Japanese process. In addition, they say, the quality of decisions is better since the people who are most knowledgeable about the matter are making the decision and the process allows workers to have a sense of empowerment in regard to their work environment.

Division of Labor

Another area of difference is that in traditional bureaucracies there is a precise division of labor with specific responsibilities assigned to each position, and each position is identified by a detailed job description. Japanese organizations assign responsibilities, defined only in general terms, to groups. The individual member's role within the group is left unspecified. The group is rewarded or sanctioned as a whole for their success in accomplishing the group tasks. This requires that each group develop internal informal processes of motivating employee participation in group tasks to augment organizational mechanisms, and that individuals be responsive to group pressures.

Job Security

Japanese organizations share certain characteristics with the bureaucratic model that sets both apart from many Western for-profit organizations. Job security is one such characteristic. In analyses of Japanese management, this is called "lifetime employment" and is frequently cited as the key element of the management style. A whole constellation of other features depends upon job security for their effectiveness. In the Japanese model employees are hired straight from school. Since they are being hired for thirty years or more and are not hired to fill a specific position (there are no detailed job descriptions or precise individual responsibilities), the criterion for hiring decisions is not technical expertise. Instead, qualities such as the ability to learn, to work with a group, loyalty, and commitment are more important. Success in school as measured by test scores and teacher recommendations is one indicator of these personal qualities. This is a partial explanation for the tremendous pressure experienced by elementary and high school students in Japan. Their performance in school is a crucial determinant of which employer will offer them a job at graduation, and this is the only opportunity most will have to get a job offer from the large prestigious organizations that practice Japanese management.

In Japanese managed firms, employees are viewed as raw material, an investment to be developed. Thus, employees are schooled in the company culture and given whatever training or education the organization might need. They are rotated throughout the various departments in order to acquire an overall view of the organization, become multi-skilled, and build relationships with other employees. Rotation and training are less extensive for employees hired from high school who fill lower-level positions and who will not be climbing to the top of the management ladder.

Organizational Culture

Like Western bureaucracies with lifetime employment (e.g., the Methodist Church, the Marines, and the Red Cross), Japanese organizations create and maintain an organizational culture. There is a sense among employees that the organization has a history, a mission, a tradition, certain ways of doing things, and an ideology. Employee orientation programs, company publications, uniforms, songs, stories, social gatherings, celebrations, and ceremonies are organizational mechanisms to sustain and build culture. One of the purposes of organizational cultures is to convince employees that the organization is, in Cole's (1979) terms, a "community of fate." Cole says that in community-of-fate organizations employees believe that they share a common destiny with each other. They will all sink or swim together. This is particularly pertinent in regard to the relationship between management and labor. In a community-of-fate organization, employees believe that they have common interests with management, and they believe that management recognizes the critical nature of their contribution to organizational success. To promote belief in the community of fate, Japanese organizations minimize the overt differences between management and labor. Managers and nonmanagement employees share a common "one employer" work experience and both are assured of lifetime employment. Both wear the same company uniform, park in the same parking lot, and eat in the same cafeteria. Managers are more visible to employees since there are no private offices or private secretaries, and manufacturing managers are expected to walk around the plant and interact with employees. Finally, there is less salary disparity between the two groups.

Lifetime employment, organizational culture, and the community of fate encourages workers to view their work as a vocation. This translates into higher employee loyalty to the organization and higher employee commitment to the achievement of organizational goals. Advocates of Japanese management contend that the ability of Japanese organizations to elicit higher employee commitment to the organization in large part accounts for the success of those firms (see Alston 1986; Dore 1973; Ouchi 1981; Pascale and Athos 1981).

Although there are some interesting variations, Japanese management is essentially similar to bureaucracies in regard to the handling of files, rules and regulations, and the practice of viewing administration as a full-time job. Since these features do not distinguish Japanese management, as it is defined in the literature, from traditional Western models of organizing, I will not elaborate these characteristics here.

Two characteristics of Japanese management that are not usually included in descriptions of bureaucracy are quite important to an understanding of Japanese organizations: *welfare corporatism,* an idea first presented by Dore (1973), and *company unions.* Welfare corporatism defines a reciprocal relationship between employees and the organization. In the ideal type Japanese organization employees define their jobs broadly, are willing to do whatever is necessary to assist in organizational success and identify with the organization's goals. In turn, the Japanese organization defines its obligation to employees in a holistic, broad manner. This means that the organization is concerned with the total welfare of the employee and provides a wide range of employee benefits including dormitories for single employees, family housing, or loans to purchase a family house, educational classes (ranging from statistical analysis to flower arranging), access to recreational facilities on site and off (mountain lodges, ocean resorts, etc.), and even assistance in finding a spouse.

Union membership is more common in Japan than in the United States, however Japanese unions are organized differently than American unions. In the United States unions are organized by trade (the United Brotherhood of Electrical Workers) or industry (The United Auto Workers). This format gives members greater power in negotiations with management and represents an alternative claim on employee loyalty. Unionization by industry necessitates that workers recognize the commonalties and interests they share with workers who are employed by other organizations including competitors. Unionization by trade goes even further and serves to divide employees within the same organizations. It suggests that workers from one skill area do not have like interests and circumstances as workers with other skills, even though both work for the same organization. These cross-cutting loyalties can lead to fissures within the organization, threatening the community of fate.

After World War II , unionization was encouraged by the American-inspired Japanese government and, as a result, union membership grew rapidly. However, the trend became less popular with American supporters and Japanese government officials when some of the unions began to join together and became increasingly militant. Citing Andrew Gordon's study of Japanese labor relations, Kenney and Florida (1993, 29) say that "the business counteroffensive was massive. The combination of management, organized under the Federation of Employers, and the state defeated the unions in a series of major strikes during 1949–1950." Japanese businesses established the current self-contained, company-sponsored unions as an alternative representative body for workers to replace the more independent, more powerful, more militant unions that emerged after World War II. These new unions take a less confrontational stance toward management. Understandably so, since managers are usually part of the membership and, in some firms, managers hold a union office as part of their management rotation. Company unions encourage and reinforce employee loyalty to the organization and do not challenge management's power, the community of fate, or other aspects of the company ideology.

Overall then, Japanese management shares with bureaucracy the characteristics of treating administration as a full-time job, positions are to be viewed by the office holder as a vocation, the existence of pyramidal hierarchy (although the shape and certain internal workings vary) and job security, a feature more likely to be shared with public bureaucracies than with private. The two differ from each other in the assignment of responsibilities, decision making, and promotion and salary determination. Finally Japanese management contains certain elements not found in the Western model: welfare corporatism, community-of-fate ideology, and company unions.

Given our belief that the bureaucratic model is the most efficient and effective way to organize human endeavors, Japanese management should not work very well. Indeed, the general consensus about it during the fifties and sixties among scholars and managers was that it was a hold-over from the not too remote Japanese feudal past. Abegglen said that the lifetime employment system was an economic liability that discouraged innovation and left organizations with the dead weight of older and superfluous workers. A respected Japan expert, Chie Nakane, was especially critical of the seniority pay and promotion system and contended that it would have to be modernized to more closely conform to the Western style if Japan was really serious about competing with American organizations. Imagine the dismay and shuffling of theories when, without changing styles, Japanese businesses became major players in the world's economy effecting what some have called an "economic miracle" in Japan. Suddenly writers, researchers, and managers began looking for the "secret" to Japanese success, and many concluded that the peculiar style of Japanese management is the primary explanation.

Still, there are significant numbers of people who maintain that Japanese companies have achieved success in spite of their organizational style, not because of it. For example, Jon Woronoff (1983) and Wolf Reitsperger (1986) assert that the successful Japanese companies owe their prosperity to their innovative methods of work structuring, plant layout and design, and steadfast attention to detail, quality, and discipline. Other analysts identify external organizational relationships as critical to Japanese management in addition to, or instead of, internal attributes. Hiroshi Okumura (1984) and Michael Yoshino (1968) point out the advantageous features of enterprise groups and advice cartels that are characteristic of Japanese organizations. Enterprise groups (often referred to by the Japanese term *keiretsu*) include suppliers, subsidiaries, a variety of similar-sized manufacturing firms, banks, trading companies, and insurance companies. The relationship a Japanese firm has with other members of its enterprise group, its suppliers and banks, is often exclusive, supportive, and noncompetitive. This facilitates low inventory, "just-in-time" manufacturing, lifetime employment (the smaller suppliers absorb employment uncertainties), and a long-term perspective, since the major sources of financing are sympathetic enterprise, group banks, and insurance firms.

Advice cartels are working relationships between government ministries, trading companies, financial agencies, and manufacturing firms for the purposes of exchanging information and coordinating policies. This arrangement has supplied invaluable marketing information to firms in their international sales operations and has provided consultation and direction concerning which product lines or industries should be promoted and which should be phased out as a part of national policy.

In addition there are other extra-organizational factors that have often been referred to as important elements in the Japanese success. Such factors as a highly disciplined, well-educated, and subservient work force; a currency that was devalued in the world market below its actual value; state-of-the-art facilities built in the aftermath of the destruction of World War II; financial and technological assistance from U.S. businesses and government; and a high degree of national homogeneity and cohesiveness have all been at some time used as explanations for the Japanese "miracle."

The Significance of the Transplants

Not until the mid 1980s, when Japanese organizations attempted to transplant their style of organizing to their foreign operations, was there any way to isolate and critically examine the effect of some of these elements upon organizational success. Previously, all the significant components of Japanese management were present only in Japanese organizations operating in Japan with Japanese employees. Earlier manufacturing, banking, and trading efforts by Japanese organizations in the United States were managed with the style common in American organizations (Pascale 1978; Lincoln, Olson, and Hanada 1978; Ouchi 1981). Even today, there are a limited number of Japanese companies that organize their overseas operations using Japanese management (Sethi, Namiki, and Swanson 1984; Milkman 1991; Abo 1994). Kenney and Florida (1993), who conducted a large-scale study of Japanese transplants in the United States, inform us that the Japanese automobile, steel, and rubber companies, latecomers to the United States, are convinced that their success depends upon Japanese management and that they can make it work with American employees.

Now for the first time there is a change in the nationality of the work force experiencing Japanese management. We can determine whether the Japanese style of organizing leads to high employee commitment (and ultimately high employee performance) or whether the extraordinary performance of Japanese employees is due to their cultural penchance for hard work (or perhaps to the inordinate amount of social and cultural dependence of Japanese workers on their employer). Put another way, will American workers respond to the Japanese style of organizing by working hard to help the organization achieve its goals? If American workers demonstrate high commitment in Japanese organizations operating in the United States and these organizations are successful, then we will have support for the contention that the way Japanese firms organize human activity is more effective than the standard Western model. There could still be a number of plausible explanations for the success of Japanese companies, such as keiretsu support of the American operations, manufacturing technology, more aggressive marketing, and so on. We will, however, have eliminated the subservient, well-educated, hard-working Japanese work force explanation.

In this context, the study of how Japanese management is enacted with American employees can be quite instructive. This book joins a growing body of scholarship focusing on the Japanese transplants in the United States and Great Britain. These studies are concerned with discovering how the internal organizational features of Japanese firms contribute to company success, posing three distinct but sometimes overlapping explanations (see figure 1). One explanation (referred to briefly above) was first elaborated by Ronald Dore (1973) in a comparative study of a British manufacturer and a Japanese manufacturer, both operating in Great Britain. Dore suggested that Japanese organizations are more successful than their Western counterparts because Japan's late entry into the industrial age encouraged them to heed the lessons taught by Western management specialists (e.g., Taylor and Deming). At the same time their history of nationalism led them to design organizations in which the core values of Japanese culture (e.g., group orientation, harmony, loyalty, respect for age and authority) were protected. Dore called the resulting style of management "welfare corporatism." I use welfare corporatism as the general rubric for the academic literature that claims Japanese organizational success is due to the fact that Japanese management empowers employees, provides job security and enrichment, and in return, employees display higher commitment to the organization and work harder to achieve organizational goals. Lincoln and Kalleberg's (1990) book is an excellent description and test of welfare corporatist principles.

At the other end of the ideological spectrum is the Toyotism school. I include Parker and Slaughter (1988) here, who call Japanese management "management by stress." Dohse, Jurgens, and Malsch (1985), among the first to use the term *Toyotism*, contend that it is superior to Fordism because it has solved "the classic problem of the resistance of the workers to placing their knowledge of production in the service of rationalization" (128). Put another way, Toyotists charge that Japanese-managed firms have evolved a diabolical system of worker control that is more effective at extracting surplus labor—including intellectual labor—from workers. Toyotists see exploitation where the welfare corporatists see empowerment. Thus, welfare corporatists describe job rotation and multi-skilling as enriching and empowering for workers. Toyotists, on the other hand, conclude that rotation and multi-skilling lead to worker exploitation in Japanese plants in that they provide the means whereby workers can fill in for injured and absent work mates. The ability to fill in for absent work mates makes possible the policy of no replacement workers, which in turn, increases the work load on the remaining workers. In research on Japanese transplants we are confronted with the unusual situation in which the same piece of evidence may support both explanations, depending on how the evidence is presented and interpreted.

The final explanation originates in the industrial engineering literature (Monden 1981, 1983; Ohno 1984; Ossola 1983). This perspective, which I call the production technology explanation, ignores the human resource issues central to the other two explanations and credits the Japanese success to their innovative production technology. Particularly lauded are the just-in-time inventory control system—so-called smart machines, "pull manufacturing," mechanisms for total quality control, plant layout and design for quick changes and retooling—attention to detail, and continuous improvement systems.

Research on Japanese organizations published in the 1970s and 1980s usually fell under one of the three headings with little overlap between them. The majority of work was of the welfare corporatist persuasion (Pascale and Athos 1981; Ouchi 1981; Cole 1979; Rohlen

Figure 1. Scholarship about Japanese Transplants in U.S. and U.K.

What is the secret of the success of Japanese organizations?

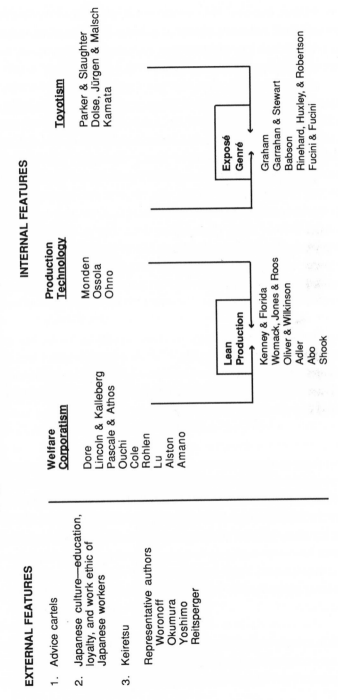

EXTERNAL FEATURES

1. Advice cartels

2. Japanese culture—education, loyalty, and work ethic of Japanese workers

3. Keiretsu

Representative authors
Woronoff
Okumura
Yoshimo
Reitsperger

INTERNAL FEATURES

Welfare Corporatism

Dore
Lincoln & Kalleberg
Pascale & Athos
Ouchi
Cole
Rohlen
Lu
Alston
Amano

Production Technology

Monden
Ossola
Ohno

Lean Production

Kenney & Florida
Womack, Jones & Roos
Oliver & Wilkinson
Adler
Abo
Shook

Toyotism

Parker & Slaughter
Dolse, Jürgen & Malsch
Kamata

Exposé Genre

Graham
Garrahan & Stewart
Babson
Rinehard, Huxley, & Robertson
Fucini & Fucini

1974; Dore 1973; Lu 1987; Alston 1986; Amano 1979; Dreyfack 1982; Gibney 1982; Hasegawa 1986). Kamata (1982) and Woronoff (1983) are among the few at that time whose analyses warn of the sinister nature of Japanese management. Both human resources explanations (welfare corporatism and Toyotism) were largely uninformed by the production technology interest of the industrial engineers, and vice versa. In contrast, the research on Japanese management published in the 1990s features a fusion of the production technology explanation with either welfare corporatism or Toyotism. While adding the production technology orientation to the human resource theories enhances the theories' ability to explain the success of Japanese managed manufacturing companies, the cost is that these newly articulated theories no longer apply to other types of organizations—for example, banks and government agencies.

Two very influential books that reflect the combination of welfare corporatism and the production technology perspective are *The Machine That Changed the World* (1990) by the MIT research team, Womack, Jones, and Roos, and Kenney and Florida's *Beyond Mass Production* (1993). Both sets of researchers contend that the Japanese auto companies (Kenney and Florida add Japanese steel and rubber companies) have created a qualitatively different manufacturing system that lowers the cost of production while at the same time increasing the quality of the product.

That system, called "lean production" by Womack, Jones, and Roos and "innovation mediated production" by Kenney and Florida, is characterized by empowered workers, innovative production technology, and familial supplier relationships. Further, these analysts contend that neither welfare corporatist human resource management, nor Japanese production technology functions effectively alone. Both must be present to realize the efficiencies and quality of Japanese auto manufacturers.

Oliver and Wilkinson (1992) elaborate this position in their study of Japanese transplants in Great Britain. They explain that Japanese lean production technology makes the organization extremely vulnerable to work stoppages, slow downs, sabotage, and so on. Therefore, it follows that successful utilization of Japanese production technology can be achieved only when workers are equally dependent upon the organization. Welfare corporatist human resource management is an internal organizational mechanism for bonding workers to the organization. Oliver and Wilkinson describe the situation in Japanese managed organizations as akin to the cold war stalemate between super powers called "mutually assured destruction."

Adler's 1992 analysis of NUMMI (New United Motor Manufacturing, Inc.), a joint GM and Toyota plant in Fremont, California, and Shook's (1988) less scientific case study of Honda in Marysville, Ohio, present a closeup view of lean production. The NUMMI analysis is particularly interesting since it examines Toyota's management of experienced UAW workers in the production of Novas (now Geo Prisms). Adler conceptualizes Toyota's management of NUMMI as a "learning bureaucracy" coupled with "democratic Taylorism." Without going into detail here, these innovations refer to Adler's claim that Toyota utilizes "scientific management," in the broad generic sense, with a participative human resource philosophy. At NUMMI, Adler reports, even the most vocal critics of Toyota's management style are in favor of the system. Their criticisms are aimed at flaws in its implementation (114).

The bulk of the new literature about Japanese transplants, however, is not heralding the innovations of Japanese management. Instead, much of it combines the production technology explanation with Toyotism, resulting in highly critical analyses of Japanese management, which I have entitled the "exposé genre" literature. Among this group is the work of Rinehart, Huxley, and Robertson (1995), a Canadian Automobile Workers team of researchers who studied the CAMI (Canadian Automobile Manufacturing, Inc.) plant that manufactures GM Geo Metros and Suzuki Sidekicks at Ingersoll, Ontario. They report a growing disenchantment with Japanese-style management among CAMI employees culminating in a five-week-long strike in 1992 after workers demanded, and eventually received, the right to elect team leaders (which has since been rescinded), a job description for team leaders, and the hiring of replacement workers for injured and absent production workers. The authors summarize their findings as indicating a growing team-based resistance at CAMI to some of management's more excessive demands and policies.

Babson (1995) and Fucini and Fucini (1990) document the discontent of UAW employees at the Mazda–Ford joint venture at Flat Rock, Michigan. There, too, the team-leader role was one source of conflict between management and workers and, as at CAMI, the union was able to negotiate the right for workers to elect and recall their team leaders. Babson believes that the presence of the UAW made possible adaptations in the Japanese style of management, team-leader elections as only one example, which have created more worker-centered work teams.

Although both CAMI and Mazda at Flat Rock are unionized, most Japanese auto transplants are not. Without union support, researcher access has been restricted to primarily superficial information supplied by management contacts. Thus, there is a dearth of "in-plant" and worker-oriented research about nonunionized transplant operations. Graham (1995) solved the problem of access by becoming a production employee at the Subaru Isuzu Automotive (SIA) facility at Lafayette, Indiana. Her analysis of the SIA production system reveals worker empowerment coupled with heightened managerial control (through the mechanisms of work standardization and low team leader-to-worker ratios). She challenges Kenney and Florida's conclusion that Japanese auto companies have successfully transferred their style of management to the United States. Particularly debunking of the community-of-fate ideology, according to Graham, are the signs of worker resistance evident on the SIA shop floor. Graham contends that researchers who lack knowledge of the day-to-day work life of employees may arrive at erroneous conclusions about the successful implementation of Japanese lean production in the United States.

In a scathing criticism of Nissan at Sunderland in Great Britain, Garrahan and Stewart (1992) charge the company not only with exploiting its British employees but also with abusing the British economy and nation. Garrahan and Steward maintain that Nissan forced the British government to provide generous subsidies for it's G.B. plant, externalized some of the risks and costs associated with auto manufacturing to local suppliers, and brought to Great Britain only low-skill, low-tech assembly jobs. They criticize aspects of Nissan's production technology for ultimately leading to worker deskilling, and for creating fear and stress among workers. Garrahan and Stewart conclude that Nissan's indoctrination of employees has been so effective that they do not know they are miserable.

Putting aside Garrahan and Stewart's inflammatory tone, their analysis offers a valuable insight into the internal dynamics of lean production organizations. They state that "at Nissan people are encouraged to view work problems as personal issues about which the company can do little" (115). Put another way, Nissan has evolved a system that encourages workers to take personal responsibility for the organizational problems of manufacturing a high-quality vehicle as efficiently as possible. Policies and procedures used by Toyota to encourage workers to take personal responsibility for cost reduction and product quality are elaborated later in this book and elsewhere (Besser 1995). However, at least in Toyota's case, this is a reciprocal relationship where the company assumes some responsibility for certain employee personal issues such as child care and personal health.

I position my research in both camps. I find that lean production as practiced at TMM is extracting more surplus labor than would occur in a traditional manufacturing plant. Production work at TMM is monotonous and mind numbing in spite of job rotation, team work, and job enlargement programs. However, those facts alone are not sufficient in my mind to condemn the lean production system, given the current economic climate. I concur with Berggren (1992) that from a pure quality-of-work life perspective, the craft teams utilized by Volvo are preferable to lean production. Nevertheless, the cost of producing an automobile this way makes the system unsustainable. Volvo's craft team manufacturing cannot compete with either Fordist producers or lean production plants. The Uddevalla plant, Volvo's showcase of the craft team system, has been reorganized since Berggren's book was published.

The research reported in this book adds to the existing scholarship in three ways. First, we will examine a variety of dimensions within the organization—for instance, the official policy, the views of critical and contented workers, managers (as official spokespeople and as employees), office staff, and Japanese managers—and show how they relate to each other. I agree with Graham that management's take on the organization's culture is incomplete, but so also is the workers' perspective. This book attempts to give voice to a broader cross section of organizational actors. In so doing, we are able to consider issues not addressed by the other research, such as the situation of injured workers, the stress of managers, the women of Toyota, and the implications of routine monotonous assembly work to the lean production system.

Second, a theoretical framework is presented here which explains how the principles of Japanese management are translated by TMM into the day-to-day actions of workers through organizational policies, procedures, and structure. I will show how the Toyota culture is recreated at TMM and how it is changed to accommodate American norms, laws, and values.

Finally, this study is unique because it is about Toyota. Mazda, Nissan, Honda, Suburu, Mitsubishi, Izuzu, Suzuki, GM at Saturn, and Ford have all, with varying degrees of success, attempted to emulate Toyota's production system. Toyota is the standard by which other car manufacturers measure themselves. Indeed, both Womack, Jones, and Roos and Kenney and Florida find a qualitative difference between Toyota's plants in quality and efficiency compared to all other car manufacturers. Even the strongest critics of Japanese management pay Toyota tribute by recasting Japanese management as Toyotism. How Toyota transplants its production system and organizational culture to the United States is important because Toyota is the elephant among the lean production manufacturers.

The purpose of this research originally was to develop a theory to explain what work is like for American workers at a Japanese organization in the United States. During the course of the study—the subsequent rereading of transcripts, data, and documents and then writing and thinking about this topic—it became obvious to me that the most important phenomena occurring at this plant is the transplantation of the Toyota team culture. Toyota's management was using a variety of policies, structures, and procedures to create and nurture the Toyota team philosophy among their U.S. employees. As I describe what work is like for American TMM employees, I will use the team concept as an organizing format, as a way to integrate, explain, and make sense of what happens at TMM. In the conclusion I will return to the bureaucratic and Japanese management models to ascertain if TMM U.S.A. conforms to our understanding of Japanese management and, if so, what implications that has for organizations.

2

METHOD OF STUDY

In the research methods section of a book, the author makes the case that the findings presented in the book are an accurate depiction of reality. Generally this is done with a detailed account of the overall research design, sample selection procedures, strategies to ensure validity of the indicators of the phenomenon, precautions against researcher or other forms of bias, and so forth. I will attend to these matters later in the chapter. First, however, because my research methodology (grounded theory methodology) may be less familiar to the reader than quantitative survey techniques, I think it is important to establish the logic of this method of discovering social reality before proceeding. I ask the reader's indulgence while I take a verbal side trip to present an amended version of Zelditch's (1970) typology of information and his strategy for ascertaining which methodology for discovering reality is most appropriate for each kind of information.

Zelditch argues that the most important criteria for selecting an appropriate research methodology are adequacy (referred to here as "validity") and efficiency. Validity is the extent to which a methodology can adequately reflect the truth of the social phenomenon under study. Many factors in a research scenario, in addition to the choice of type of methodology, influence the validity of the results. Those other elements are put aside for this examination. The question concerning us here is, Will the methodology yield valid results under the best of circumstances?

Efficiency is the ability of a methodology to gather information with a minimum of effort, expense, and waste. Two methodologies may be able to generate equally valid information. The preferred one is the technique that requires the least expenditure of time, energy, and other resources. Why survey ten thousand people when the data that would be generated already exists? Why survey one hundred people when an interview with one will do? The most important criteria of the two is validity. Efficiency is irrelevant if the methodology cannot generate a true picture of social reality. The ideal is a method of knowing that is both valid and efficient.

Validity and efficiency are the yardsticks used here as we consider the different types of information and which methodology is most appropriate for each. Let us turn now to Zelditch's first kind of information—frequency distributions or the count of social phenomenon. There are several techniques that can generate valid frequency distribution information: surveys of a probabilistic sample or the population, in-depth interviews, and participant or nonparticipant observation, to name a few. As an example, if we are interested in knowing how many auto workers in the United States are satisfied with their jobs, we could uncover that information (assuming we used a valid measure) by surveying the whole population of auto workers, by surveying a random sample of them, by conducting in-depth interviews of the population or a random sample or, alternatively, a team of researchers could spend years observing, interviewing, and interacting with workers in auto plants. The time and energy required to observe and interact with auto workers in sufficient numbers to allow for a valid count of satisfied workers is extremely prohibitive. Even conducting in-depth interviews with enough people to learn the frequency of job satisfaction among auto workers is very costly, requiring a huge team of researchers. Thus, using the gauges of validity and efficiency, the best methodology for gathering frequency distribution information is, in most cases, a survey of a probabilistic sample. Interviews and participant observation can be both valid and efficient methods for gathering frequency distribution information for a small group—for example, a street corner gang or a family—but not for large groups like auto workers.

The second type of information is concerned with the meaning, for individuals, of events, relationships, social structures, roles, norms, organizations, and so on. Several researchers and authors have considered this type of information in regard to the subject of work. Terkel's (1974) examination of what work means to people, Kamata's (1982) account of life in a Japanese auto factory, Rohlen's (1974) anthropological analysis of work in a Japanese bank, and Hamper's (1991) tales from an American auto assembly line are classic examples of meaning information.

Due to the nature of this kind of information, there is no easy check of validity. True meaning is whatever credible members say it is. How many members should be polled to get a valid picture of meaning for the whole group? Glazer and Strauss (1967) suggest the use of what they call "theoretical saturation" to determine the sample size necessary to understand and explain the meaning of social situations. When a researcher interviews members of a specific group about the meaning of some aspect of reality and the same responses, themes, concerns, and feelings come up again and again—so much so that the researcher can predict the interviewee's response—then theoretical saturation has been reached on this issue for this group. Nothing new is being learned on this subject from interviewees and therefore, according to Glazer and Strauss, a sufficient sample of this group has been gathered. The researcher might choose to go on and ask this question to members of comparison groups to delineate and elaborate the meaning of events learned from the first group.

Consider an example of theoretical saturation from this current research. One question put forth was, What advice would you give a friend who was considering coming to work for Toyota? After fourteen of the first fifteen interviewees indicated that they would tell their friend not to come to work for this company unless they were really willing to work hard, I decided that theoretical saturation had been attained on this question. This does not mean that I concluded that everyone at the plant held these same opinions. I could draw no conclusions about the frequency of these opinions based on a nonrandom sample of fifteen. Since the "working hard" response was volunteered (the question was not even directly related to amount of work), since all respondents mentioned working hard—either negatively or positively in the context of other questions—and since the same theme appeared in company publications and quotes from Toyota officials in newspaper articles, I felt it was safe to conclude that "working hard" was a cogent part of the work reality of workers at TMM. If I had encountered workers who said that working for Toyota was easy, a piece of cake, it would not have disproven the significance of "working hard." Instead it would have provided an opportunity to elaborate the "working hard" belief. Why would some workers believe in it and others not? How do the believers differ from the nonbelievers? The task would be to determine the meaning and significance of the differences, the relationship of the differences to other emergent themes, and use the opportunity of the anomaly to further elaborate the meaning of "hard work" to TMM employees.

An essential element in research pertaining to the meaning of social reality is clear specification of the reference group to which the meaning applies. In the case above, I needed to learn whether the group with the belief that "working for Toyota is hard work" consisted of all workers in the factory, only production workers, only management, workers with no manufacturing background, or others. The researcher uses comparative analysis to discover and specify the group involved and then limits generalizations to that group on that particular aspect of meaning. Thus, Hamper can only provide insight into the meaning of work to line factory workers in an American auto plant during the post–World War II era. Kamata can only generalize his conclusions to line workers at Toyota auto body plant in Japan in the 1970s.

The methodologies commonly used to count phenomena—surveys and compilations of official data—are not of themselves antithetical to ascertaining valid meaning information. If allowed to, I could have given workers at TMM a questionnaire with an open-ended question asking them to give their advice to friends about employment at this plant. From their responses I would look for consistent themes, concerns, and such, in the same way that the written transcripts of my interviews were analyzed. I would undoubtedly have discovered the importance of the belief in working hard, but I would have responses from many more people than was needed to establish theoretical saturation. Surveys of large samples to gain meaning information may be valid but are inefficient. Additionally, they do not allow the flexibility to adapt and change questions as themes and concerns became apparent. Surveys of small samples using open-ended questions are really not much different from face-to-face interviews, at least in regard to the focus in this analysis.

The final category of information is general knowledge information that covers expectations about the social structure, norms, laws, processes, and policies of a particular group, organization, institution, or society. Examples of this type of information are the accepted form of marriage in the United States, the expected class load of faculty at the University of Kentucky, the time first-shift workers are expected to be at their stations at the line at TMM, the rate of hourly pay of production workers, and the prerequisites for the team-leader position. The best way of discovering this type of information is to find a credible member of the group under study and ask him or her. Any normal adult in the United States can inform us that the preferred form of marriage in this country is monogamy. Any production worker can provide information about starting times, break times, and pay rates at his or her company. The gauge of the validity of the answers regarding general knowledge information is the credibility of the informant: Is the informant telling us the truth and does the informant have access to the knowledge we seek? If both of these questions can be answered affirmatively, a sample size of one is sufficient to provide valid general knowledge information. If we doubt the credibility of our informant, we may ask the same question of another credible person. However, once we have established the credibility of our informants, it is unnecessary and, therefore, inefficient to repeat the question with a larger number of people. The validity of our information is determined by the credibility of the informant and has no relation to the size of the sample. Under the right circumstances, then, interviewing an N of one is sufficient to establish validity and is the most efficient manner of gathering general knowledge information.

My research questions (How is Japanese management being transplanted to TMM and what does it mean to the American workers involved?) pertain to meaning and general knowledge information. In part, emphasis on these two kinds of information resulted from my personal interest in the topics, but also because I was unable to gather valid frequency distribution information. Toyota was not willing to allow me official research access to their operations. Therefore, given my limited resources, it was impossible for me to draw the probabilistic sample of workers necessary for frequency distributions. An ideal case study would contain all three kinds of information. Nonetheless, much can be learned from meaning and general knowledge information if the researcher can establish the credibility of his or her informants, the logic of the analysis, and so on. That brings me to the specifics of my research procedures.

The Research Design of This Study

The scientific procedure used in this research—grounded theory—is a technique of systematically organizing and interpreting observations.

After selecting a subject of interest, the first step in the process of grounded theory is theoretical sampling. This is a process whereby the analyst simultaneously collects, codes, and analyzes his or her data. The data used in grounded theory can take any form. The researcher may interview, informally talk to, and/or interact with some of the participants in the social situation. Statistics from surveys, census data, and other sources can be used. Available documents and publications are studied. Personal observation and even personal experience in the situation are acceptable forms of data. Strauss comments that "experiential data are essential data, because they not only give added theoretical sensitivity but provide a wealth of provisional suggestions for making comparisons, finding variations, and sampling widely on theoretical grounds. All of that helps the researcher eventually to formulate a conceptually dense and carefully ordered theory" (1987, 11). The form of the data is not critical, the processes used to analyze the data are, however. The researcher must immerse himself or herself in the data, going over it systematically, meticulously, and repeatedly.

Grounded theory employs a concept called "slices of data" to utilize as many different sources of data as possible regarding a subject. Each of these kinds of data—quantitative data, document analysis, personal experience, historical analysis, crosscultural comparisons, biographical accounts, field observation, and open-ended interview—is capable of providing a validity test for the others. Additionally, contrasting slices of data should force the analyst to integrate and expand the theory to account for contradictory or inexplicable data. The result of using these various modes of knowing is a theory that is expanded, enriched, more clearly delineated, and verified to a certain degree.

In my research of TMM, I utilized five sources of data. First, I examined company publications intended for customers (found at any Toyota dealership), employees (the employee handbook and a magazine published monthly and mailed to employees), letters sent to employees, and company annual reports. Second, local and national newspaper articles proved to be quite useful, especially during the first couple of years of TMM's U.S. operations. Both of these sources provided coverage of employee demographics, injury rates, official policy statements, and so forth. In addition, the company publications contained articles about internal organizational structures and policies. Since they were used by TMM to help propagate the company philosophy, these publications provided insight into the nature of that philosophy.

Third, I conducted a line-by-line analysis of over one thousand pages of typed transcriptions of thirty-six in-depth and focused interviews with TMM employees. These employees were from a wide variety of departments and hierarchical levels. They were both Japanese and American, male and female, injured and healthy, satisfied and dissatisfied with TMM. I purposively selected diverse categories of interviewees so that they would provide me with multiple comparative samples. However, I had at least two interviewees in every category and eight who were rank-and-file employees. I was able to interview six employees from the same department, four production workers, one skilled worker, and a first-line supervisor. Seven departments were represented among my interviewees.

I started with a set of open-ended questions, some of which changed when it became apparent that certain questions were not appropriate or did not generate very informative responses (see Appendix A for a copy of the questions that were asked of all interviewees). At the same time that I was interviewing employees, Toyota publications and transcripts from former interviewees were also being analyzed. If analysis revealed that clarification was needed on a topic or resulted in formulation of an hypothesis that needed verification, I would select a category of employee who could respond to my query, arrange an interview with someone in that category, ask the arranged set of questions and also the new questions that emerged from the analyses. In this way the interview questions evolved over the course of the research.

Among the employees interviewed were relatives of students, neighbors, members of my church, friends of friends, and others I knew in the Georgetown community. These people supplied the names of other employees later contacted for interviews. Only one person refused to be interviewed, saying that he did not have the time. Another never showed up for scheduled meetings. All in all, TMM employees were very cooperative and generally seemed to be quite open and frank about their experiences at Toyota. To counter the possibility that all the people interviewed might be "of like mind" about their experience with TMM, interviewees were asked to suggest people whose attitudes differed from theirs who might be willing to be interviewed. I assured all interviewees and informants that their names and identifying characteristics would be kept confidential. The names used in this book are pseudonyms.

A fourth source of data was my own personal experience with Toyota over a five-year period extending from 1987 (prior to the plant opening) to 1992. I had regular access to TMM family functions through a close relative who was a TMM manager. I attended all the major social events, which included plantwide happenings such as yearly company picnics and perfect attendance parties; departmental functions such as happy hours, Christmas parties, and summer picnics; social functions for the management staff; and nonofficial functions such as company-sponsored whitewater rafting trips, athletic events, and fishing excursions. I toured the facility twice. It was not until 1989, when I officially began this research, that I started recording interviews and making field notes after conversations and events. I had numerous contacts with Toyota employees—both prior to this time and during the research—that were not recorded in any form. They did contribute nonetheless to my understanding of and impressions about Toyota.

I often used informal conversations with employees to ask questions for clarification or to test hypotheses. For example, after analyzing the transcripts of interviews with several employees, I realized that "on restriction" was an important issue at TMM. I also realized that I did not understand many of the details about the policies surrounding the label of *on restriction*. There were several employees with whom I interacted regularly and this was a general knowledge subject they were very happy to explain to me. I was also able to go back to two interviewees whenever necessary for elaboration, clarification, or to see if my interpretation of events and issues corresponded with theirs.

Finally, the extensive literature on the subject of Japanese management and the material written specifically about Toyota were utilized. These sources gave an understanding of how Toyota operates in Japan and provided in-depth information about the Toyota production system, its uniqueness and history. Eisenhardt (1989) points out that prior research and literature about a subject can provide a valuable source for comparative analysis, stretching and validating the theory emerging from other sources of data.

3

THE CONTEXT

Before we consider the Team Toyota culture at Toyota's Georgetown operations, I will set the stage, so to speak. It is important to know some concrete facts about Toyota, TMM in Georgetown, Toyota's American employees, the physical facilities, and the usual work day in order to understand the more theoretical dimensions of work in a Japanese transplant. Toyota was founded in 1937, is still a family-controlled company and has its major Japanese facilities and its headquarters in rural Aichi Prefecture. *Fortune* magazine (July 1993) reports that Toyota is the largest manufacturer in Japan, the fifth largest industrial organization in the world, and the third largest automobile producer in the world. Compared to other Japanese automobile manufacturers, Toyota is a latecomer to the American manufacturing scene. Its first foray into the United States was in 1984 at NUMMI (New United Motor Manufacturing, Inc.), a Toyota–GM joint venture. It was not until December 1985 that Toyota announced plans to build its first U.S. assembly plant in Georgetown, Kentucky, to manufacture Camrys.

Initial employment at TMM leveled off at approximately three thousand employees producing 240,000 Camrys annually. An engine plant was added in 1989 bringing the number of employees up to about forty-five hundred. Construction of additional assembly capacity, completed in 1994, allows Toyota to double the number of Camrys manufactured and provide room for the production of the new Avalon (the redesigned Cressida). This was expected to increase employment to approximately six thousand employees. Since start-up of assembly operations, the presence of Japanese employees has declined significantly. Each department continues to have a Japanese coordinator, and/or executive coordinator, the president is Japanese and a small number of Japanese trainers spend three months at Georgetown and then return to Japan.

Toyota's Georgetown operations are nonunion. They utilize extensive screening and training of their American employees including assimilation in company ideology and manufacturing processes and trips to Toyota's Japanese operations for employees at first-line supervisor level and above. According to Toyota's Japanese executives interviewed, Toyota believes in the uniqueness of Japanese management and regards the major tenets of that management style as transferable to their U.S. operations.

Other significant organizational features involve the famous Toyota manufacturing technologies developed or refined in Japan. These include just-in-time (JIT) manufacturing processes (JIT is a whole manufacturing system based on the premise of minimum inventory of stock—parts and products, usually requiring deliveries from suppliers daily or several times a day, "just in time" to be used in the assembly process) and the kanban inventory control system; jidoka, which requires the line to be shut down by automated machines and/or line employees whenever a defect or problem is spotted; standardized work; and kaizen, the philosophy of constant improvement. These processes will be elaborated upon only insofar as they impinge upon workers' day-to-day actualization of the Toyota system.

TMM's American Workers

The following statistics about TMM's workers appeared in an article in the *Lexington Herald Leader* (Prather 1989). Only 4.4 percent of those who applied for work at TMM were actually hired. This allowed Toyota to, as Toyota's vice president of human resources said, "have the luxury of hiring the cream of the crop." TMM's workers are well educated: 13 percent have graduated from college, 60 percent have some postsecondary education, and 27 percent have only a high school diploma or its equivalent. The workers are relatively young by industry standards, with the average age at thirty-two years, and are inexperienced in manufacturing work. 13 percent of the work force is composed of minorities compared to 8 percent minorities in the local population, and 25 percent are women. 95.4 percent of employees are natives of Kentucky. There was a concern during negotiations to bring Toyota to Kentucky that Toyota might import a large share of its labor force from other states and/or from Japan and not provide significant employment opportunities for in-state workers. As a result of these concerns, Toyota agreed to hire a high proportion of locals in its work force.

TMM was described by several of my informants as possessing a flat organizational hierarchy. That is to say, it was their opinion that TMM had few layers of management above the line workers and a low proportion of the total work force in management and support roles. The statistics support the employees' opinion, indicating that 67 percent of employees are line workers, who are called "team members" at Toyota; 14 percent of all employees are team leaders; another 4 percent are group leaders; and 14 percent of employees are involved in administration. The team-leader position is not intended to be management. The pay differential for team leader is only 5 percent above team members and the team leader does not officially evaluate team-member performance. The team leader can be more appropriately described as fulfilling the lead-worker role. Including the team leaders with the team members, 81 percent of employees are directly involved in production activities.

The 14 percent of employees classified as administrative are subdivided into the categories of clerical positions, called assistant and associate staff; specialists who are professionals, such as accountants, engineers, purchasing agents, and human resources professionals; and upper-level management. It is in the capacity of upper-level management that most Japanese employees of TMM serve. Japanese employees also function as trainers, but these are temporary positions with the trainers generally staying in the United States only three to six months or rotating with other Japanese trainers on a three- to six-month basis. The number of trainers has greatly decreased since start-up in 1988. Japanese nationals who are in management positions usually stay in the United States for three years, but some have stayed longer. This is not to imply that the majority of management positions are filled by Japanese nationals. They are not. However, all departments are staffed with either a Japanese manager or one or more Japanese coordinators to assist and advise the American managers.

A Normal Day

After asking each interviewee about their background and the process they had to go through to get their job with Toyota, I asked them to tell me what they did in a normal day. This seemed to be a difficult question for workers to answer. Either they talked about all sorts of things related to the day—assuming, I think, that I could not possibly be interested in what they actually did—or their days varied so much that there really was no "normal day." Production team members fell in the first category, with specialists and managers in the latter. After much prodding they all did eventually answer my question. As I have implied, however, the patterns vary according to work category. I will concentrate on the team-member position, finishing by briefly contrasting other categories of workers to the team member.

Production Workers

Production work cannot be understood without first visualizing the facility in which the work is conducted. The manufacturing plant is immense, with acres under one roof. My impression was the same as that of Janis Gibbons, a production team member. She called it fascinating, awe inspiring. The stamping machines are huge, several stories high. In other areas there are machines upon machines that to the unschooled viewer appear to need no human intervention. The line itself weaves a contorted path through the facility, sometimes running at waist level, sometimes overhead, sometimes with other lines feeding into it. Adding to the busy, complicated picture are miles of pipes, belts, and conduit running through the plant appearing to come and go everywhere. What seemed particularly remarkable to me was the control, power, coordination, and planning that the machinery, the plant layout, and the utilities represented. Janis Gibbons, a production worker, expressed a similar reaction when she said, "Before you work in a place like this, you never think about what goes into making an automobile. But now that I'm here, I am amazed at all that's involved. I am fascinated by it. I love to go out and just watch the machinery run."

Compared to other manufacturing plants I have visited, the Toyota plant is clean and uncluttered. This is partially due to the newness of the plant, but also to the meticulous Japanese concern for cleanliness and putting parts and tools back in their proper place after use and to the JIT system that does not allow a buildup of parts or inventory. Bicycles are available at key locations to facilitate travel by anyone needing to get from one end of the place to the other. Andon boards, prominently displaced everywhere in production, are arrayed with green, yellow, and red lights intended to alert maintenance staff and management to problem areas on the production line.

Most of the plant consists of the assembly area, with subsidiary functions physically adjoining assembly. Administrative functions are performed in a separate part of the facility, but production management is located in the production area. There are no private offices, only common offices sometimes called "pits" by the employees who work in them. Private meeting rooms are available if privacy is required. Space is set aside in a few areas for pingpong tables and small basketball courts. TV monitors, vending machines, tables, and chairs dot the facility in what are called "break areas," where restrooms and lockers are also available. Apparently the Japanese designers of this facility did not understand what it meant, in terms of restrooms, to have a large contingent of female employees. One common complaint among female team members is that sometimes they must return to the line without the opportunity to use the restroom during the break because there just are not enough of them. This has since been addressed by management and women's restrooms have been added.

All employees, regardless of position, are given a similar uniform of beige pants and navy blue shirt. As one person told me, "They all look like gas station attendants." However, employees are only encouraged, not forced to wear the uniform. In production, deviations usually consist of employees wearing a Toyota t-shirt of some kind and jeans. Because the uniform pants do not fit female employees very well, they are more likely to wear jeans and, in assistant staff positions, to wear more fashionable street clothes. The uniform is topped off with a Toyota cap decorated with chevrons to indicate team-leader and group-leader status. Caps are not required and, again, are more likely to be worn by males than females.

The first shift of workers arrives about 6:15 A.M. on weekdays. Employees don whatever safety clothing they are required to wear, which may include coveralls, glasses, hard hats, gloves, wrist braces, or steel-toed shoes. Toyota insists that the line begin at exactly 6:30 with everyone in place. Therefore, it is implicitly required that workers arrive at least five minutes early in order to be in place on time. Most team members come earlier than that, with some arriving more than thirty minutes before the line starts. John Perkins comes in early to "shoot the breeze" with the other members of his group, to make sure all his parts and tools are in place and ready to go, and to put on his safety glasses and gloves. In departments where more extensive safety clothing must be worn, workers have to arrive around thirty minutes prior to start-up. Workers are not paid for time put in before the line starts.

From 6:30 until the break most team members repeat the same job approximately every sixty seconds. If the job requires parts or tools, team members are responsible, with the assistance of team leaders, for seeing that those parts and tools are available and accessible. The Toyota goal is constant work (100 percent) with no breathers during the two hours or two hours and fifteen minutes between breaks. However, this is a goal that is seldom achieved. With the jidoka production system, the line always stops, sometimes several times a shift, sometimes for many minutes at a time. To the team members this means that when the line is running, the pace allows no rest and even replacing parts becomes difficult, but there will be periodic breaks in effort when the line stops. If the stoppage occurs because of a problem in the team member's area, she or he must assist in doing whatever is necessary to get the line going again. If the line stops because of a problem elsewhere, team members use the time to restock their parts, clean the work area, or visit with each other. According to Barbara Woulton, a production team member,

> If you're running 80 percent or 88 percent, that's no problem. You start running up around 92 percent or 94 percent, you're going to be working pretty constantly and it will tire you out. But you get used to it and you can make yourself work that hard. 94 percent that's really good. That's the most they can expect. Most the time we never get up to that.

There is no policy against team members talking to each other during the performance of their job. However, many team members do not have the opportunity for socializing because of the noise in their area, because their job is too far from other team members for normal conversation, or because they are simply too busy to talk.

The first break occurs at 8:45. Five minutes of the break will be used by the group leader to review yesterday's performance with the group or in a safety meeting. Some departments are more lax in conducting these meetings than others. Team members use the restroom, get a drink or a snack with the remaining time, and head back to the line. A bell rings at five minutes before the line starts to alert workers to be ready to begin work precisely when the break is over. If they are on a team that rotates, they will now be performing a different job than before the break.

Work continues until 11:00 A.M. when the line shuts down for the forty-five-minute lunch break. Lunch can be eaten in one of the Toyota cafeterias, the group break room, or team members can run out and pick up food at a neighboring McDonalds or other restaurants. Sometimes pizza or other food is ordered in as a PT (personal touch) function (more will be said about this later); that is, Toyota foots the bill for lunch. Lunch break has become the occasion for pingpong matches, basketball games, card playing, and other recreational activities. One team member says pingpong and basketball allow him to "expend energy during lunch" so that he is ready to go again at 11:45 when the line starts back up. In some groups, group leaders designate a certain day every month when the team meets alone for lunch—paid for by Toyota—in a private meeting room to talk about team concerns. At 11:40, the bell rings warning workers that they have five minutes to "clean up your lunch, get your tools lined up, and get back out on the line by 11:45." Again, if they are rotating, they will be working at their third job of the day.

Afternoon break comes at 1:45. Many groups have another safety or quality meeting lasting five minutes and then do whatever they like for the ten minutes remaining. By afternoon break, team members in most departments are informed about how much overtime will be required that day. This allows them to use break time to call anyone who needs to know about their quitting time. Overtime is the norm. Team members I interviewed expect it and look forward to the paychecks that reflect it. The exact amount of overtime that is required on any given day is not known, by management or team members, prior to the afternoon break. Even without being told by management, team members can generally estimate how much will be needed by early afternoon. The amount of overtime is based on the production quota and the amount of line stoppage that has occurred during the shift. I will not elaborate how this system works here, but will return to it later in the main text of this analysis. If overtime is to be an hour or longer, team members would take another five-minute break at 3:15 before going into overtime. Generally overtime lasts half an hour to an hour. That means the usual quitting time is between 3:45 and 4:30 and the normal work day is 6:15 to 4:00. Some production departments must stay longer than this to clean up, and in some areas quitting time is more independent of the production schedule and, thus, more difficult for team members to predict.

Second shift starts at 5:15 P.M. and ends at 2:00 A.M. The beginning of the second shift at 5:15 limits the amount of overtime the first shift can work to one and one-half hours. Correspondingly, there is a larger window of time between the ending of the second shift and the beginning of the first shift at 6:30, allowing second shift to work more overtime "if they ever needed us to." Typically the second shift has between one-half hour and one hour overtime, also. But one second-shift team member reported working one and one-half hours of overtime for three weeks straight. The normal work day is 5:00 P.M. to 2:45 A.M. for second shift.

Production team members do not work Saturdays or Sundays. Toyota policy discourages it, although this is not to say that it never occurs. John Perkins told me of a letter from TMM's president Mr. Cho sent to all team members elaborating Toyota's policy on overtime and apologizing to the spouses and families of team members for the amount of overtime recently required. In the letter Mr. Cho stated that Toyota did not want team members to work on weekends and would try to avoid it because Toyota recognized that weekends were for families.

Skilled Team Members

Team members who belong to what are considered the skilled trades are primarily employed in maintenance positions at Toyota. Their usual day is quite different from production team members. Maintenance team members also arrive early to "shoot the breeze" with the team members on the third shift, read the paper, or drink coffee in the break area. Maintenance staff work three shifts compared to the two shifts of production work. At 6:30 they meet with the third-shift team leader, go over any problems that occurred during the third shift, discuss what needs to be done, and generally exchange information. After that they "strap on their belts" and head out to examine the line. Each maintenance worker is assigned a particular area of the line for normal maintenance, to monitor and troubleshoot. But the team member must be available to assist when problems occur in another part of the line if she or he is needed. Their first break occurs at 8:30 also, but frequently they work during the break:

> If something breaks and we think we can fix it in fifteen minutes, we'll go ahead and work on it during the break. Then when the line starts, if everything is going okay, we take our break. (Raphael Martinez)

If there are no problems with equipment, the maintenance team members will work on kaizens (suggestions for improving the production process) or other projects assigned them by their team leader. They get together with production team members (as they work on the line or during their breaks), with other maintenance staff, team or group leaders, and perhaps with safety and engineering specialists when appropriate, to brainstorm and come up with suggestions for improvement. Then the maintenance team members build or modify equipment per the suggestions. This continues until about 11:00 or 12:00, when lunch break occurs. The maintenance staff lunch break is flexible because it may be necessary for them to work on line equipment when the line is shut down for lunch. If skilled team members are not working during lunch, they will eat in the break area or cafeteria with production team members.

The duties of maintenance staff are much the same in the afternoon as in the morning. Sometime during the day, reports about problems, repairs, kaizens, and so forth, must be prepared. Maintenance staff, at least in some departments, meet with vendors and make recommendations about equipment purchases. At 1:45 they have the afternoon fifteen-minute break. The second-shift maintenance staff starts at 2:15, but since they generally come in early, they will be arriving around the afternoon break time for first shift. First-shift staff may socialize informally with second-shift personnel if there is nothing pressing at the moment. At approximately 2:15 they meet with second-shift team leaders to update them on what occurred on first shift. At 3:15 first shift is over. Maintenance staff do not work overtime during the week but usually make up for it by working Saturdays and holidays to perform the maintenance on machines that can only be done when the line is down. Unless there is an emergency, team members leave work at 3:15. Keep in mind that maintenance staff are always on call during their shift to respond to team members who are having a problem with a machine or who think everything is not as it should be, and to any line shut down in their area. I revealed my ignorance by asking Martinez why it is that something is constantly breaking down and/or in need of continual maintenance when Toyota has a new facility with new machinery and conscientious employees. He said some of these machines perform the same function—stretch the same belt, bend the same joint, and so forth—thousands of times a day. It is only to be expected that normal wear and tear will cause them to need oiling or replacement of hoses and belts, and that it is to be expected that things will go wrong at least once or twice in ten thousand performances of the same operation. He tried to make it simple for me by adding, "It is sort of like machines getting repetitive motion illness just like people do who have to perform the same motion over and over again." In addition to the daily tasks, all team members in most departments—production and skilled—"get out the brooms and mops and clean the entire area" on a weekly basis. According to Martinez , "It makes it nice working in an area where people put up their stuff and things are kept clean."

Maintenance staff are frequently involved in some kind of training program. A maintenance group leader, Roger Slider, described to me in detail his plans to upgrade the skills of his maintenance group by involving all of them in various training programs. He uses the overlapping time between shifts for thirty-minute or forty-five-minute daily instructions in some aspect of the operation. Sometimes the format is that the electricians teach the other team members about electrical procedures, or a team member skilled in a particular machine or function will be the teacher. Occasionally, one or two team members have taken a formal training course and will relay the information back to teammates. Ralph Canty, another maintenance team member, recalls that he has been involved in a training program of some kind for almost the whole two and one-half years of his employment with Toyota. A great deal of the training takes place on Toyota time. Canty was participating in a program during the time of our interview that required three hours of his time after the 3:15 quitting time every day for four weeks. He was not required to do this, but looked upon it as an opportunity. Martinez speaks positively about the training opportunities he has had with Toyota, especially insofar as they have contributed to broadening his skills. He is an electrician and says that at other places where he has worked "an electrician only did electrical work, millwrights had their area, pipe fitters had their area, and none of them ever crossed over. Whereas at the Toyota, maintenance people are trained to fix whatever breaks down, whether it is electrical, hydraulic, pneumatic, or whatever. If the problem requires you to communicate with an engineer, you'll do that. If a vendor has to be contacted, you'll do that too."

The normal day of the skilled team members at the Toyota is varied and challenging and ends promptly at 3:15. No wonder interviewees in other categories of positions, including specialists and group leaders, seemed to envy the skilled team members. Eric Klaske, a production group leader, said "if I could be born again at Toyota, I would want to come back as a maintenance team member." The two skilled team members I interviewed each had opportunities to become team leaders but refused to leave their current role because they enjoyed it so much.

Managers and Specialists

The work of managers and specialists, like that of skilled team members, is varied and challenging—perhaps too challenging. The managers and specialists I talked to reported working a minimum of sixty to sixty-five hours per week, including some weekends and holidays. When asked how work at Toyota had affected other areas of his life, Ron Prichard, a group leader, said,

> I didn't have anything else. There was nothing else. I would work twelve hours, come home and eat, read the paper, and fall asleep. I used to wake up in the middle of the night and think, *I didn't do that,* or *What am I going to do about this?* . . . One time I worked three weekends in a row, both Saturday and Sunday. You have to keep your nose to the grindstone to the point where the job completely dominates your life. . . . There were mornings when I walked in to work and thought, *Wasn't I just here? Where's the rest of my life?*

Prichard has since left Toyota. He said that several other managers have left also, because "we just don't need or want to work that hard." Another manager Eric Klaske reported,

> It's not the fact that I work twelve to thirteen hours a day, busting my butt, that bothers me. What bothers me is that I work that long and I still need to stay another two to three hours to get caught up. I'm working that hard and I'm not getting everything done. I'm not being a good group leader to my people, because I don't have time for them.

Later he provided a poignant example of the kind of stress Toyota managers often experience:

> Nothing is accepted as being good enough. There was one day that we had all kinds of problems. I felt like myself, the team leaders, and the team had done a marvelous job of handling everything that happened to us that day. It was like coming under attack and living through it. I felt like we had just been bombarded and everybody had reacted well. Everybody pitched in and gave 110 percent and we made it. The next morning I came in to work feeling great, like I had really accomplished something. Two or three Japanese trainers sat me down and pointed out everything I had done wrong the day before. I fell apart. I started crying. I just couldn't take it because I felt like I had done a real good job. We all pulled together and got through this and the next day they unloaded on us.

As Prichard described it, the drive for constant improvement is felt most strongly by managers. He said,

> I felt that no matter what I ever did, it would never, ever be enough at Toyota. There was never, ever a point that I felt I could reach where I was on top of my job. What I'm talking about is reaching a point where you know you've accomplished something, that you're in control of the job and that

you don't need to keep your nose to the grindstone to the point where the job completely dominates your life.

Moreover, managers and specialists do not receive in return for their extra time and effort the usual perks enjoyed by management at many organizations: no special cafeteria, no special parking lot, no private secretaries or private offices, and little flexibility in scheduling work hours.

Conclusion

An appropriate way to summarize what a normal workday at TMM is like is to relate some of the interviewee responses to the question about what advice they would give someone who was considering working for Toyota. Prichard said,

> You'll be expected to kill yourself for Toyota and you'll be paid a decent living wage. If you can take it, you'll be taken care of for the rest of your life. When they say you start at 6:30, they don't mean 6:31. You start at 6:30 and you kill yourself for two hours, take a ten-minute break and kill yourself for two more hours. So if someone wants to do that in exchange for the money and security, go for it.

An assistant staff, Nancy Henderson said,

> I think it takes a very special type of person to work in the Toyota office environment because of the open office setting—it being so hectic and stressful. Plus, the volume of work that we do is so great. A person has to be able to handle the stress, work well under pressure, and be willing to work hard, do whatever is necessary.

According to Pat Griffin, an engineer,

> Basically, if they are the kind of person who is dedicated and wants to work hard, wants to be rewarded, and wants good benefits and good job security, then I would recommend that they go to work for Toyota. But if they're not willing to give 100 percent, I think they will be really miserable here.

Doug Huntington, a manager, advised,

> I think there's a good future with Toyota. There's tremen-
> dous opportunity and stable employment. But the person
> should realize that he is going to have to give 100 percent all
> the time, and if he is not ready to do that, then Toyota is not
> the place for him.

A production team member Dave Fetalucci has recommended Toyota
to several relatives and friends as a good employer:

> You have to work hard here, but Toyota is a good employer.

Team member Tom Lemming responded,

> I would tell them that there's a lot of really good benefits
> there and if they do end up in the production area—and the
> vast majority of the people do—you just can't get that pay
> anyplace else. But I've made the statement before, its the
> closest thing to slave labor you'll ever see, as far as the
> actual work goes. The workers just grow accustomed to that
> hyper style of work.

Team member John Perkins would recommend Toyota to anybody
except people with bad attitudes. When asked what he meant by "bad
attitudes," he explained,

> A bad attitude is someone who has a very narrow mind.
> They're there just to do their job. They don't want to put
> forth any input into any type of group discussions—just an
> overall anti-social type person. We don't need that. At all.
> Because Toyota's success is due to employees being com-
> mitted and participating.

Whether positive or negative, every respondent counseled that
people who were not willing to work hard should look elsewhere for
employment. They recognized that working for Toyota is hard work in
a disciplined format, requiring extraordinary commitment and effort.

4

TEAM TOYOTA

Overview

The concept of "team" is central to understanding the internal functioning of Toyota at its Georgetown, Kentucky, plant. It is a primary element in the Toyota philosophy, an integral part of workers' daily interactions with each other and with the organization as a whole, and related in one way or another to all other concepts that will be discussed in this study. So much of the vocabulary, philosophy, and structure of the organization are oriented around *team* that it must be regarded as the central theme of the Toyota culture, even if its realization appears to be more a goal than an actuality in some work groups and some departments.

Specifically, consider how the importance of team is acknowledged in TMM's policies, structures, and vocabulary. In *Toyota Today* (Fall 1988), the publication for customers, a manager tells how the team philosophy is communicated to employees as he describes the hiring process for new employees: "By the time candidates get through the process, they realize it's not just a single job they're applying for, but membership in a team." The theme is repeated in every aspect of the employees' interaction with the organization. Put another way, the hiring process is only the beginning of the employees' introduction to the team concept. A significant portion of the orientation for new members deals with the concept of team spirit and cooperation.

All employees who are involved directly in production are formally—and even informally, in daily conversation—referred to as "team members." Each team is made up of four or five peers (team members). This is the primary functioning unit of the production and maintenance processes. Their direct supervisor—or, perhaps "lead worker" is a more appropriate term for this level—is called the "team leader." A cluster of three or four teams is referred to as a "group" and is headed by a group leader. The group and the group leader's responsibilities are defined in relation to the teams supervised. Each level of management is defined in relation to the functions of the teams and team members that constitute the department, with the exception of administrative operations. The publication sent to all employees, *Toyota Topics*, stresses the importance of team. One regularly appearing feature is entitled "Team Players." It highlights various employees and presents a short biographical sketch of each. Another feature published intermittently is entitled "Team Spirit." In addition to the emphasis on team in the publications, the company store is called the TEAM store and is loaded with t-shirts and other clothing carrying team slogans.

Clearly, the team concept is officially promulgated by the organization. But is anybody buying it? How effective has the organization been in convincing the employees that TMM is a big team and they are team members? One theme that occurred consistently in the interviews with the employees was the concept of team. Although the interview questions were quite open ended, it is notable that, under these circumstances, workers mentioned the significance of team without being asked or prompted. Usually team was referred to explicitly by team members. Barbara Woulton remarked, "The team work concept really does work here." Bob Dottard said, "There's a lot more emphasis on team work than any other place I've worked. Instead of turning people against each other, they try to get people to rely on each other." According to John Perkins, "The management is very open to suggestion. Everybody is a team member trying to achieve a common goal." At other times the reference was more oblique. A manager, Doug Huntington, commented, "They feel that whenever possible, we need to get the team members, team leaders, and group leaders involved with the decision." Still, in other instances, the importance of team was affirmed by the interviewee, lamenting its inadequate realization in his or her work area: "I don't believe the Japanese got what they intended, if this is what they intended—for everyone to work together and work at things" (Janis Gibbons, production team member).

Team: Definition

For the purposes of this analysis, team is defined as "people who work together cooperatively for the common good." There are three types of teams mentioned by workers and referred to in Toyota publications. The first is the small work unit, the "work team." Production and maintenance functions are organized so that tasks are assigned to these teams. Ideally, all members of the team know how to do all the jobs assigned to the team. Members of these teams know each other personally and socialize with each other in nonwork settings. They could be described, using Cooley's terminology, as a "primary group."

The second type of team, which I will call the "company team," expands the unit of analysis and includes all employees of the local operation. In this sense, all TMM employees, regardless of position in the hierarchy or function in the company, or education and prior experience, are viewed as members of the same team, sharing the same goal. The purpose of the company team is to provide the context and the ideology for intragroup cooperation and commonality of goals among all employees. The members of this team do not know all other members personally. Nonetheless, TMM encourages friendly rather than formal relationships among members.

The third type of team follows logically from the other two, but was only referred to explicitly by interviewees who were managers and/ or Japanese and in Toyota publications. This team, which will be called the "corporate team," is at still one higher level of analysis. It includes all members of the Toyota corporation, including, but not limited to, manufacturers in the United States and Japan, their suppliers, various other subsidiaries, and the semi-independent marketing and sales corporations affiliated with the corporation. For those employees who have visited Japan, where the majority of the corporate team presides, this team is more of a reality than for other employees. The employees who have visited Japan are also likely to be in positions of management that would allow them the scope to see the corporate linkages. One manager, Doug Huntington, discussed how the management of this larger team (the corporate management) could see the big picture that includes markets, financing, and other corporate members inaccessible to local managers of TMM. Further, the corporate team leaders make decisions that can affect the environment of the local company, such as offering incentive plans for buyers. The relationship between the corporate leaders (who happen to be Japanese) and the local American managers was described as similar to a teacher-student relationship by Huntington. The existence of this corporate entity is further recognized by some interviewees through the systems and operations adopted locally that were initiated and developed by the corporation in Japan. Returning to the Cooley dichotomy of primary and secondary relations, the corporate team is characterized almost totally by secondary relations, at least for the American employees. But the use of words like *family* and *team* by the corporation to describe itself in the annual report, employee publications, and consumer publications reveals the desire to appear—and the importance of appearing—like a warm primary group.

The Work Team

How do you know a work team when you see one and once you have one, how do you maintain it? I will attempt to answer these questions by elucidating the properties of work teams as they exist at the TMM. Work-team membership means working cooperatively with other team members. One interviewee, when asked to sum up what it was like to work in a Japanese organization, expressed it by saying, "I'd really hate to say what is the most important. Maybe the team atmosphere. Maybe I didn't emphasize that like I should have. The team atmosphere is I think, a really good way to work. We really use it in our team. We help each other out tremendously. One person gets behind, I don't mind a bit going to help. I think that's pretty much true throughout the plant" (Dave Fetalucci, team member).

Going beyond cooperation, team membership implies that other team members are the primary appraisers of one's work. In a way, they, not management, are the consumer and the evaluator of most importance. Woulton elaborated,

> If I don't do my job properly, the next person on the line is going to know about it and they're going to tell me about it. For that reason I want to do my job better. I want to do a good job. I know that that person is coming behind me. They're going to have to do something after me and if I don't do my job right I've messed them up. I know I don't like to be treated that way.

Later, she added,

> Also, if you make a mistake and it gets to inspection, every-
> body's going to know you made a mistake. Nobody wants
> to be pointed out in the group: "You made the mistake
> inspection caught." It happens and we all understand
> that. . . .You don't want people to know you make mistakes.
> You want everybody to know you did a good job.

Ron Prichard, a former production supervisor, contrasts this with
another manufacturing facility at which he now works:

> The atmosphere of cooperation, you know, let's make the
> top. Let's make as many cars as we can and make them as
> good as we can, that's the prevailing atmosphere at TMM.
> Whereas at my current employer, the prevailing atmosphere
> is, "I don't care. I don't care if everything's not right. No
> one will ever know." Quality doesn't matter. Production
> doesn't matter.

The amount of management control and the difference in the values and
norms of the small work groups at these two facilities is striking.
Researchers like Burawoy (1979) have documented other examples of
work-group cultures that are completely distinct from and contrary to
their company philosophy. The difference between the attitudes at
TMM and Prichard's current employer did not occur by accident. As I
will point out later, the cooperativeness among team members and the
congruence of team values with company values are the result of well-
orchestrated policies and procedures.

Another property of the work team is that members have relationships with each other that go beyond the work roles they perform. This
is certainly not unique to TMM. What is notable is the extent to which
Toyota encourages warm holistic relationships between work-team
members. The small size of the work teams facilitates the possibility
that members can get to know each other quite well. Money is made
available to group leaders and other supervisors to assist in paying for
social functions for the group. This is called PT (personal touch) money
and the functions so sponsored are PT functions. Group leaders may
decide to distribute part of these funds for team social functions. However, PT functions are usually organized at the level of the group or

department and include Christmas parties, family picnics in the summer, bowling nights, trips to baseball games, or pizza ordered in for lunch at work. Furthermore, to assist in building social relationships among work teams, TMM policy discourages breaking up teams in work assignments unless it cannot be avoided. This might occur if someone in the group leaves, gets promoted or transferred, or because teams become unbalanced within a group. If for example, due to the loss of team members, one team has three inexperienced workers and another team with no loss of personnel has all experienced workers, then the teams may be broken up and rebalanced to avoid an overload of work on the one experienced member of the first team.

Work teams have a structure. All team members are viewed as peers by TMM and each other. Ideally, so that job rotation can occur, each team member can perform all the jobs assigned to his or her team. They all receive equal pay. The only pay distinction is the lower entry-level pay that increases at six-month intervals until at eighteen months the peak of the pay scale for team members is reached. Team members in the skilled trades do have a different pay scale than other team members and reach their peak wages after three years of related experience.

Team leaders receive a 5 percent increase in wages over team members. This position is supposed to be filled from the ranks of the team members supervised or—since there are two teams performing each function, one team for each work shift—the position may be filled by a team member on the same team who is working a different shift. In fact, this is more likely to be the situation. TMM policy is that team members desiring to be promoted to team leader must begin as team leader on second shift, allowing second-shift team leaders to fill the first-shift vacancies. Of course, at start-up it was not possible to have team leaders who were promoted from the ranks, and under certain other circumstances this is not possible. Team members who desire to be promoted to team leader must take a series of three courses. Each course consists of two hours after work for two weeks on the employee's own time. Therefore, if there is no one in the team who has taken these classes, a team member from another team, preferably within the same group, will be chosen.

The responsibilities of the team leader include setting up the work schedule, responding to the andon signal (this signal indicates that a smart machine or a production employee has spotted a problem on the line; if the problem is not immediately remedied, the line will be stopped), replacing parts and supplies of team members, and filling in for team members who are absent or on restriction. (A complete chapter will be devoted to "on restriction." Briefly, employees on restriction are unable to perform their normal functions due to a work-related disability.) They may also be assigned special projects by the group leader. Often, team leaders serve as first appraisers of team members' ideas for work improvement suggestions ("kaizens") and the person team members turn to when they have problems with the company. It is due to the nature of the team leader's job that prior experience as a team member is so desirable. The team leader does not evaluate employee performance, at least not officially. This is the group leader's responsibility. Nor does the team leader act as an official conduit for upper-level management communication to employees. This too falls within the purview of the group leader. It appears to be Toyota's goal for this position to be viewed by team members as an experienced lead worker, willing to pitch in, get his or her hands dirty, and do whatever is necessary to assist the team in performing its function. Clearly, by providing such a small pay differential and denying the team leader evaluative duties, Toyota is signaling that this position is not a member of management. Even so, the position is highly sought after by team members because it is believed to be a stepping stone to management levels and, in some work areas, allows more variety of work and more opportunities for intellectual challenges.

The team members I interviewed saw the team leader as a quasi supervisor, usually fulfilling the role described above. In one group, team members portrayed widely differing perceptions of the team-leader role. But even Gibbons who was most dissatisfied, was conversant with the Toyota model and believes that her dissatisfaction results from the model's inadequate implementation in her work area. Gibbons commented about team leaders:

> I feel like there's still a hangover from American management. What the Japanese intended and what we got are not the same things. What we were told in assimilation is not what we got, particularly in the very beginning

More specifically, Gibbons added,

I do see a lot of people who are reluctant to jump in there and do the physical labor and it's not just within my group, it's in other areas. They aren't there. They earn 5 percent more, but they aren't there to do the labor. They're there to oversee the labor.

The team leader is a critical position for articulating between the work team and the company team and in actualizing the company-team concept to team members. I will return to this aspect of the team-leader position in the section on company team.

The Creation of Work Teams

The Employee Screening Process. Employee screening and orientation are critical elements that contribute greatly to an atmosphere conducive to work teams and, as I will demonstrate later, to the company team as well. The large number of team members hired initially for the first shift and then, later, for the second shift went through the same basic procedure. Interviewees tell me that the process has been abbreviated for more recent hires, but none of them is sure what the changes are. For the first groups (the majority of current team members) the process was to fill out an application form and take a half-day aptitude test administered by the state employment service. If an applicant made it through this screening, he or she next completed two days of hands-on and situational tests designed to assess (according to interviewees conclusions) physical dexterity, ability to work with a group, problem-solving skills, motivation, and creativity. An interview with a potential supervisor and a human resources representative followed for applicants successful in the second phase of screening. A physical exam precedes or follows the interview. The whole process, from application to hiring, usually requires a minimum of six months.

What are the results of such an arduous application process and how could Toyota get enough people to submit to it? Let me begin with the last aspect of this question first. At full employment Toyota has hired only 1.4 percent of those who have made inquiries about employment (Prather 1989, B5). Obviously Toyota had no problem attracting interest among potential employees. However, it is unknown outside TMM how many of those inquiries culminated in a completed application. Certainly not all of them—maybe not even a majority. Still, the number is large enough that almost all interviewees knew at least one person who took all the tests but did not get hired. I believe we can understand why these people were willing to go through this long process by considering interviewees' responses to the question I asked them regarding why they came to work for TMM. All team members mentioned the relatively high salary offered by Toyota as a prime motivation. Many of them expressed how their increased wages since coming with Toyota have changed their lives. Team member Gibbons says, "This is the first time in my life that I've had enough money to take care of myself and it makes me different." According to team member Fetalucci, "After my son was born I had, at one point, five jobs. Now I do my job. I come home. I spend more time with my kids. I've gotten close to them since then. Because I never saw them before. I was gone from dawn to dark."

Examining the educational characteristics of the team members might also illuminate this issue. Thirteen percent of team members have college degrees, 60 percent have some postsecondary education, and 27 percent have a high school diploma or equivalent (Prather 1989, B5) What is truly remarkable is that TMM hired 95 percent of its employees from Kentucky, where the approximate high school dropout rate is 50 percent and was able to hire people with postsecondary education for 73 percent of the team member slots. Why would a college graduate go to work on the line for Toyota? John Perkins answers, "The pay here is really good. More than I've ever made. I love it. I have a college degree and I was having trouble finding $7 per hour jobs. It's just ridiculous."

While money may have been the primary reason people would go through Toyota's application procedure, other factors played a part as well. These include opportunity for advancement, security, wanting to be part of something new and exciting, and a fascination with Japanese management. Fetalucci was the first to bring this to my attention:

> I feel like I'm part of something that is really coming up. I feel like I'm part of a really good, respected company, a company that people talk about. Whereas before people didn't talk about my employer, my job. I have more identity in this situation. Everybody is interested in what Toyota is doing and people ask things about it. People do interviews with me about my job.

There were negative factors (lack of sufficient equivalent opportunity in the area) and positive factors (wanting to be part of something important and interesting) involved in motivating people to apply and follow through with the application procedure.

Returning to the first question posed at the beginning of this section, what are the results of this application procedure? Toyota generated a great deal of information about each applicant upon which to make employment decisions. Whether those decisions were better as a consequence cannot be answered with the data available for this analysis. Nonetheless, some things clearly are evident from this process. First, the length of time it takes to complete the process is itself a test of motivation and perseverance. Second, the fact that so much time and energy is expended by Toyota in this process communicates to potential workers that Toyota thinks the ordinary workers are important for overall company success. Third, the process communicates to employees that they and other team members were chosen on the basis of ability and potential as measured in an "objective" way. This would encourage the belief that jobs were not given to people on the basis of prior experience or whom they know. In this respect it constitutes a method of articulating components of the Toyota philosophy to employees. It fosters respect and a certain esprit de corps among team members. Each team member knows that every other team member had to submit to the same strenuous entrance requirements and passed muster.

Fourth, those applicants who make it through screening and become employed might be expected to feel exceptional, as if they are among the chosen few. One employee is quoted in *Toyota Today* (Fall 1988, 17): "Just making it through the testing and assessment process gives me a sense of pride." I expected this sentiment among the interviewees. All of them were asked about the process they had gone through to get employed. However, none of them expressed this emotion. They were pleased to be employed. But they were not proud in the way expected of someone who had gained entry to an elite group. One cannot generalize from the relatively small group I interviewed to the whole group of team members. Still the total absence of this emotion causes speculation. I believe the disparity between expectations and observations has to do with the educational background of those team members interviewed. All had some postsecondary education, including two who had graduated from college. Each seemed to be proud to be a member of the company team but not particularly proud to be at the team-member level. The ramifications of this will be explored later.

Finally, if Toyota is even remotely successful in being able to determine which employees will be effective in working in groups and hires only those applicants who demonstrate this skill, then the work-team concept has a better chance of realization. Even if the group skills measurement instrument is not very valid, the application process itself supplements the screening, provides messages about Toyota philosophy to new employees, and sets the stage for feelings of equity and esprit de corps among team members.

There is one more aspect of the screening process that must be addressed before we move on to other antecedents of work teams. It is a fact that most of the team members do not have prior manufacturing experience. Whether this is an intended consequence of screening or a function of the background of applicants from this geographical area is unknown and, for our purposes, unimportant. What is important is that this was intended Toyota policy however achieved, and the reason for this policy is that inexperienced employees are as close to a blank slate in work habits and attitudes as Toyota could get. In Toyota's view they have developed no "negative" attitudes that must be overcome. The benefits of this policy to TMM and to the company's facilitation of the work-team concept are made clear by Prichard, the former group leader:

> You couldn't take some of these old boys I'm working with now and take them into TMM and expect Toyota to ever hire them. He (the worker) wouldn't get it. If the shift is over at 12:00 midnight, about 10:30, they say "We've done a good day's work," and they don't do anything else the rest of the day, and that's common. . . . If anybody wanted to spend ten, twenty years there—or thought they might have to spend ten, twenty years there—you can bet they'd want to fit in and do like everybody else, and why not. It's a lot easier.

Additionally, since the team members are unlikely to have prior manufacturing experience, the majority of them have never been affiliated with a union. This, too, appears to be an intended outcome of the screening process.

Prichard summarized the effect of Toyota's hiring policies: "They've got literally the most intelligent, the most driven, and the most conscientious and cooperative people that were available to work for them. They've skimmed off the top, the cream."

It is important to note that in this discussion of screening my focus has been on team members and not managers, assistant staff, or specialists. The procedure is different for them because these employees are not hired to work in work teams.

Assimilation. Immediately following hiring, team members are exposed to a week of assimilation (orientation) to the company. Assimilation represents the employee's first introduction to what it means to be a member of the work team and the company team. Approximately half the week is spent in the classroom and the other half on the line. Team members are taught introductory problem-solving techniques, team work concepts, aspects of the production system, and the overall Toyota philosophy in the classroom. On the line they become familiar with the specific tasks they will be performing and, in addition, they see a live demonstration of the communication skills, team work, and philosophy as they are enacted by other team members and team leaders. This assimilation scenario is more likely to be true for the employees hired in the large waves of staffing the first and second shifts. Since then, the process has been shortened in some departments to place new employees on the line as quickly as possible.

As a result of instruction in the philosophy of team work and actual experience with it, many employees become firm believers in its benefits. My research cannot say how extensive these attitudes are among team members. Instead I note that almost all team members interviewed mentioned it positively. I have already quoted some of their comments. One team member Tom Lemming, who recognized several problems with TMM, was sold on the team concept. He attempted to convert me, emphasizing how the pursuit of knowledge was not being promoted by having individuals, instead of teams, working on dissertations. He said,

> Now if you worked at it on a team level and three or four of
> you had this paper to do and if all of you generated and went
> out and talked to people like myself and all of you got the

same grade, then you would get a lot more information and
a lot more accomplished.

Lemming considers the major obstacle Americans see to the team
method as he continued,

> But, you're also going to have that one or two weak people,
> or two or three people are not going to do half the work that
> you do, and that's going to create hard feelings and you're
> going to feel, why should they get the same grade I get when
> I did twice the work. But that's just the way that we have
> been taught through the years. Whereas, the Japanese phi-
> losophy is not that way. It doesn't make any difference if
> you're put together on a team and you see one person is not
> doing as much or pulling his weight. You approach that per-
> son and say you know he's got a problem. Maybe, they're
> just a slower person, maybe they're just weaker mentally.
> So you try to find what their strong points are. Maybe their
> strong point is not interviewing. Maybe their strong point is
> in the library digging up information. That's the way they
> approach it.

To illustrate that this philosophy is actualized, recall Woulton's statement, quoted earlier, that if she does not do her work correctly, some
other team member down the line will come and tell her about it.
According to the philosophy, as the team members are taught it, this
feedback from one team member to another is not placing blame, it is
identifying a problem and attempting to find a solution.

Despite screening and assimilation, some teams are going to run
into the problem of slacking members and subsequent animosity
between teammates. Introduction to team concepts and techniques of
problem solving in assimilation provides a tool for teams to use in dealing with recalcitrant members when they are encountered. Universal
familiarity among team members with problem-solving techniques
facilitates and legitimates team members addressing these issues in the
open, in an informal, nonthreatening way. If problem-solving strategies
are followed according to instructions, the work team is less likely to
be splintered by hard feelings and defensiveness resulting from

attempts to place blame. In addition, whether or not this tool actually works, becoming familiar with the tool and the problem it is intended to ameliorate, will forestall the anger and disillusionment that could cause team members to give up on the work-team concept when difficulties are encountered. Team members are informed at the outset that team work requires work and that problems are to be expected. This warning helps to prepare them for the probable strains they will encounter in team work later on.

Beyond the initial assimilation, employees at the level of team leader and above spend several weeks in Japan at TMM's sister plant. This is an expensive undertaking. The fact that Toyota is willing to expend this kind of money teaching employees about the philosophy of work teams and the Toyota production system underscores the importance of these subjects to Toyota corporate management. In the *Lexington Herald Leader,* a Toyota official explained the rationale for extensive employee training in Japan:

> We can't expect people to comprehend or understand the magnitude of the system or how effectively it works without them seeing it. So they have the opportunity to see it first-hand because they actually work on the production line, and they actually get to build products and they actually get to see how they would be handled as a team member. They come back charged up and ready to build quality products. (Advertising Supplement, *Lexington Herald Leader,* Oct. 5, 1988, 5)

The "they" referred to in the above quote are not the ones physically building "quality products." The key role of the team leader is illustrated in that it is through the team leader that the lessons learned in Japan will be most directly communicated to the work-team members. It is through the instruction given by the team leader, the stories told by the team leader about Japan, and how the team leader responds to various aspects of the production process—such as his or her attitudes toward work improvement suggestions, responding to the andon cord, and organizing social activities—that work-team concepts become part of the day-to-day operations of TMM. Furthermore, the team leader's knowledge of the work process and philosophy is augmented and his or her role as lead worker and trainer is legitimated. Added to assimilation and a trip to Japan, the team leaders are required to take more in-depth classes on problem solving, job instruction training, and group dynamics (how to conduct meetings, communication skills, conflict resolution, etc.) to assist them in their role as lead worker. Interwoven into the classes are threads of work-team philosophy. Dottard who attended the classes, elaborated:

> In that problem-solving class I took, they identify the problem and look for the countermeasure. Like if you're doing the job wrong, instead of saying "You idiot, why don't you do it right?" they look for the countermeasure or what's the problem there. Why are you doing it wrong? Is it because you're not trained, or the tool is too heavy, the hose is too short? They look for things like that.

Since these classes are open to team members (they must be taken by the team member prior to promotion to team leader), and team members actually take them, their importance is far greater than if only team leaders had taken them. Through these classes team members get reinforcement in the official team work philosophy at a later time, perhaps even years, after their initial assimilation.

Work Freedom and Responsibility. Screening and training are important antecedents of work teams. However, without policies and structural features that support work-team functioning, the philosophy would be only rhetoric and would quickly result in disillusioned team members. Two organizational features mentioned previously, the policies of equal pay for all team members and job rotation within the team, contribute to feelings of equality and cooperation among team members. Further, work teams are given a certain degree of job freedom and corresponding responsibility for the tasks assigned to them. Toyota prides itself on standardized work. This means that each production job is minutely described and must be religiously performed according to specifications. Therefore, when I discuss job freedom and responsibility, I am placing it within the context of a standardized assembly process. Team members are not free to perform their job a certain way on one day, change it the next, and then change back again because they feel like it. Changes can and do occur, and are officially encouraged, but only as a result of following specified procedures to gain management approval. Nonetheless, every team member and group leader interviewed had at least one story about some change that had occurred as a result of a team's initiative in identifying and solving problems. In fact, the changes were so numerous, that I concluded that work changes were the norm instead of the exception. Fetalucci affirmed this theme:

> When I first hired in, every week we had to tape lines, tape your L10 lines down. We were spending hundreds of dollars a week plant wide to tape these lines. They were using a lot of tape. I wondered why don't they paint these and leave them. But there are so many things that change, they know they can't do that because they're going to have paint there then. I see now why. We've had changes and there would have been painted lines there that shouldn't have been there. Because we change the lines a lot. Things are changing constantly.

Taping lines instead of painting them indicates that change is not only policy in the form of words but is also reflected in concrete referents and practices. In other words, it is institutionalized. The institutionalized processes to encourage teams to take responsibility for and change tasks, equipment and work environment are kaizens. Perkins described how TMM encourages kaizens and team member participation:

> That's part of being a team member, to help solve the problem. Whether it be parts problems or ergonomics (this refers to machine-human interface) problems where your hands are hurting because you're shooting a bolt. . . . The team members in my team and the group leader interacted throughout several months trying to think of a new way to redesign this building table to suit the needs of our bodies, rather than our bodies suit the needs of the table. This is basically called the kaizen, constant improvement system. That's one of the reasons for Toyota's success. They have incorporated that into their overall plan and they offer little

rewards such as gift certificates for your input into improving your overall working conditions or quality of the product, or making the process easier by changing certain things. I've already made $400 on this system.

The logic in awarding gift certificates at a discount department store instead of money is explained by Raphael Martinez, a skilled team member:

> The reason for the gift certificate, they thought about giving money on your paycheck, but they learned that whatever you got you could share with your family, bring the family in on it. My wife spends all my kaizens. She keeps saying, come up with these ideas. So that's where I got my watch from and that's where we get a lot of things. But it's nice. It works out good. So we keep trying every day to think of something that will make a certain process better.

Gift certificates necessitate the purchase of a tangible item that continually reminds the team member and his or her family of the rewards for kaizens in a way that money spent on a few beers and a meal, or put in the bank, would not. Team members not only receive gift certificates for their kaizens but are also recognized in competitions and ceremonies as well as in *Toyota Topics*.

Some kaizens require approval of other departments and upper management and may need assistance from maintenance or engineering to implement. In these instances, Dottard noted that "you just have to persevere through the system." However, other kaizens can be implemented with minimal input or approval outside the team. Woulton related: "That's what I did the other day on my lunch break. I needed a stand for one of my kaizens and I ran up and down the line until I found what I needed. I asked them if I could have it and they gave it to me."

Janis Gibbons is dissatisfied with the slowness of the system. Even so she provided an example of team freedom:

> So that afternoon I saw Charles and I said, "Charles sit down here. I can't draw. Will you draw me up a jig" (a small hand-held tool). I said, "I'll write on here what we want if you'll

draw it up and sign it because I'm not a team leader. I can't sign it." He drew a little picture of what I said and wrote down on there "as light as possible," and they made us some very, very, light jigs. They're so much easier on your hands.

This is not intended to communicate that all work changes originate with team members or within the team. Perhaps the majority do not. As group leader Prichard said when commenting on the source of the kaizens in his area:

> With Toyota's emphasis on consensus, that's really kind of blurring in my mind as to where they originated. . . . But the truth of the matter is, that I think that 90 percent of the emphasis of what was done came from me as well as the majority of the ideas for what to do. I think only the group leader is really in a position to see the big picture as far as the group goes.

Again Prichard is able to provide a contrast between TMM and the other manufacturing facilities where he has worked on the issue of job responsibility and change. In regard to the other facilities he said,

> If you turn over the reins to them (the line workers), and say, "Okay make suggestions. What would you do?" either, number one, they don't come up with anything; number two, they come up with the stupidest, dumb things you've ever heard; or number three, they just want to goof off all day. Also, management in some of these places is often inflexible, totally unwilling to try anything different—"This is the way we have always done it."

For the purposes of nurturing work teams, the actual proportion of work changes that originate with team members is not really important. What is important is that team members feel empowered to improve and change their work, are rewarded directly for doing so, and are given the freedom to implement some of the changes themselves, even if sometimes their suggestions have to run what appears to be a bureaucratic gauntlet before approval and implementation. The significance of the work team is also recognized in that their input is supposed to be sought before changes, which originated elsewhere and which directly affect them, are implemented.

To summarize this section, the organizational features that were described here as being antecedent to work teams and supporting the continuing existence of work teams were screening and training of employees; task allocation to work teams including a certain amount of job freedom and responsibility for the work team; equal pay and job rotation among team members; and official encouragement of warm holistic relationships between team members supported by small team size, relative stability of teams, and money made available to groups and teams for social events.

The Consequences of Work Team

I have divided my examination of consequences into three parts: consequences for organizational goal achievement, consequences for the individual team member, and problems encountered with the implementation of work teams at TMM. A cautious note needs to be presented before I proceed. At the time of this analysis, TMM had only been operational for approximately five years. Therefore, any conclusions regarding consequences may be premature. Generalization beyond the present time frame is ill advised. Even with these limitations in mind, there is value in understanding the consequences as they were manifested then.

Organizational Goal Achievement. Relating work teams to organizational goal achievement requires knowing what the organizational goals are. This is more problematic than it may at first appear. To avoid diverging into an analysis of the goals of TMM, about which I have gathered little information, I will use the goals TMM communicates to workers and to the public. The plant manager is paraphrased in *Toyota Today* (Fall 1988, 9) as saying, "The system (which includes teamwork, job flexibility, employee suggestions, continual training and improvement, and high quality) was developed to achieve (simultaneously) three main objectives: to produce the highest quality product at the lowest cost with a minimum of waste."

These three objectives will be used in this discussion as TMM's goals. How then do work teams contribute to the realization of these goals? The work team represents the formalization of the informal, primary group. Toyota has harnessed the power of primary relationships to motivate active team-member participation in the interests of product quality and cost and waste reduction, and at the same time, has created a superior mechanism to monitor and control team-member activities on the job. I have demonstrated previously how team norms and the threat of team sanctions encourage team members to cooperate with each other, help each other out, work as hard as they can, and do the best they can. Team approval and respect is gained when team members do a good job, or perhaps more appropriately, when they do not do a bad job. A bad job would elicit "problem-solving comments," or minimally negative thoughts, from other team members who see that "your mistake was caught in inspection," or who experience an increase in their work load as they attempt to re-do or make up for your mistake, or who must daily face feedback from the group leader regarding the quality and efficiency of yesterday's performance. Work teams, as they have been described in this analysis, will encourage workers to perform their best, to give 100 percent, to "bust their butts" as Prichard calls it.

Three examples will demonstrate how the norms and sanctions of work teams are actualized in the day-to-day behavior of team members in ways that contribute to organizational goal achievement. Team members show up for work fifteen to thirty minutes before starting time, so that their parts will be in place, any equipment they need will be located, safety geer and/or special clothing will be donned, and pleasantries will be exchanged with other group members. In this way, when the line starts they are ready to go. Team members are not paid for this time, nor are they forced by official TMM policy to be early. Official policy states that workers must be on the line exactly at starting time. The policy has an implicit expectation that workers will come in early, because coming in early is usually necessary in order to be prepared to start exactly on time. However, coming in fifteen to thirty minutes early, when five would do in some departments, has evolved because team members enjoy socializing with each other, because some are motivated to put in extra effort and time to ensure that the job is done right, and because it has emerged as a group norm. Naturally, not all team members conform to this pattern and some who do, dislike it. This is especially true in those departments where workers feel compelled to show up at least thirty minutes prior to work time. Still, it was common among interviewees to come in earlier than necessary and to express positive sentiments about it. The willingness of team members to contribute time for which they are not paid to the goal of starting the line at full speed exactly on time is the most cost-effective way of accomplishing this goal. Clearly, this is an important contribution to organizational goal achievement.

Motivation to help meet the goals of cost and waste reduction and quality are also exemplified by one team member as he described his job to me:

> From the moment I start, I'm building a gizmo or putting it on the car, I'm always looking for defective parts on the gizmo or the car. If there is a scratch on the gizmo, and I built the whole thing, I'm going to immediately ring that andon. The team leader's going to come and he's going to build another gizmo. If we notice any other problems on any other processes before us, such as scratches on the panels, bolts laying around in the trunk, we immediately grab them and dispose of them properly. Because if you let it keep on going, you'd have a little bolt in the back that would rattle and drive the customer crazy. (John Perkins)

The constant monitoring of the quality of the product that is indicated here is a fairly inexpensive and more effective method of ensuring high quality than hiring a legion of quality checkers. TMM does employ people whose only function is quality checking, however far fewer of them are necessary when team members feel that quality monitoring is part of their responsibility also.

The third example points out the impact of work teams and kaizens in problem-solving endeavors. Woulton described to me how she passed the time in what was basically a boring job:

> That's what I do while I'm working, I think of new ways to improve the area that I'm in. I just recently redid the whole job since I've been there. It took a lot of time and effort on my part, but I wanted to show them I could do it. It saved all

kinds of time as far as labor is concerned. Now my next
project will be a safety task that I'm working on.

Woulton has received monetary awards for these ideas through the kaizen process. The accumulated benefit to TMM's goal achievement of
team members passing their days in this fashion must be significant,
certainly more than the monetary awards involved.

Each of these examples demonstrates that my interviewees define
their role in the company flexibly. They are willing to perform tasks
beyond their job description, even without pay, to contribute to organizational goal achievement. The reasons they are willing to do so are
more complex than the dynamics of the small work team. The most
sophisticated screening and training program in the world could not
sustain the motivation of the work team to strive for organizational
goals without the philosophy and policies inherent in the company
team. More will be said about this in the section on the company team.

I have examined some of the ways that work teams contribute to
organizational goal achievement through motivating team members.
Let me now turn to the second aspect of this: the work team as a superior
mechanism of monitoring and controlling workers. I have already
shown how team members can control each other. What I am referring
to here is the role of the team leader. The team leader is not really management and is not a team member. As most first-line supervisors, this
position is caught between the two, management and workers. Since,
however, the team leader is supposed to be promoted from the ranks and
is not officially viewed as management, she or he may more closely
identify with team members. Minimally, the team leader should know
the language of the team members, be familiar with their concerns, and
be respected by them. He or she then could be an excellent conduit of
information from the team members to management.

On the other hand, it is not uncommon for individuals in the in-
between role to identify not with the group they came from but with the
group they aspire to join. If one assumes that the majority of team leaders aspires to move up the management ranks, then they can be viewed
as shifting their reference group from the team members to the managers. Aiding in this process, the prepromotion classes refamiliarize them
with Toyota philosophy. The trip to Japan "fires them up" to implement
that philosophy. Consider also that team members selected to become

team leaders are likely to be exemplary models of Toyota's philosophy in the first place. Through the selection process, the classes, and the trip to Japan, a shift in reference group for team leaders from team members to management becomes a logical outcome and is in fact facilitated by the procedures just described. Thus, the team leader is uniquely selected, positioned, and schooled by Toyota to proselytize the philosophy, goals, and procedures to team members in the team members' own language and in terms that are important to them.

The team leader is, ideally, known personally by team members. He or she is seen as a benevolent lead worker, a respected member of the team. Combine this with the low team-leader to team-member ratio and we have the formula for close, unobtrusive observation and monitoring of team members. Indeed, when team members believed their team leader was properly fulfilling his or her role they did not see him or her as a supervisor at all. Perkins said, "They'll walk by, they'll look but you don't feel all nervous. You say, 'Hey how are you doing.' If they want to watch that's fine." Gibbon's major complaint was that team leaders in her section were acting like overseers instead of lead workers. Put another way, team members who had a good relationship with their team leaders saw them as the embodiment of what Japanese management stood for. Those who were disappointed by team leader attitudes and performance had trouble looking beyond this person to find good qualities at any level of management. The team leader is the face of Toyota to many team members.

The team leader as a structural quality of the work team has other important consequences for goal achievement. The team leader fills in for any team member who is missing or on restriction, thus allowing the organization to operate with a leaner work force than it would otherwise be able to do. Additionally, since the ratio of team leaders to team members is high, there are more opportunities to reward by promotion team members who perform well according to TMM's evaluation. The likelihood of direct, concrete, positive reinforcement in the form of promotions to team members who show up earlier than necessary, who are constantly monitoring quality, and who contribute ideas for improvement are far greater than if the ratio of supervisor to workers were one to twenty-five or thirty, as occurs in some factories.

Team leaders contribute to TMM's goal achievement through their ability to provide information to management about team member interests and concerns. Their identification with management provides ideally a "true believer" in the Toyota philosophy to every four or five team members. The low ratio of team leaders to team members allows for close, unobtrusive monitoring and control of team members, a lean work force, and more opportunities to reward meritorious team members.

Theoretically, there is the possibility that work teams could be detrimental to organizational goal achievement. Promotion of close relationships between team and group members could lead to attitudes conceptualized in the terminology "in-group out-group." An in-group is a close knit clique that erects boundaries between itself and the rest of the world—the out-group (Sumner 1940). This is usually accompanied by feelings of exclusivity and superiority among members of the in-group as they compare themselves to the out-group. Within an organization this translates into competition, rivalry, and even hostility between different work groups, professional groups, departments, or shifts. In fact, there have been some symptoms of rivalry and hostility between departments and shifts at TMM. Some work tasks have been left purposefully unspecified and ambiguous to allow flexibility in task performance. Such things as preventative maintenance on equipment, work area clean up, and stocking of parts and tools are tasks that are required of maintenance and production work teams, respectively, but are to be performed whenever they can be worked in. When a shift or department experiences an especially hectic day, these responsibilities may be left for the next shift or team to perform. Under circumstances of the give and take of cooperation between groups, this is no problem. But if rivalry develops between groups or if it is perceived that one shift or department or team is not carrying their weight, then hostility will escalate between the groups. Intragroup rivalry within an organization is not exclusive to organizations that function with work teams. However, frequent communication and closeness between members of the same team and group may facilitate intergroup squabbles.

Just as members of the same work team could view the other teams as rivals, "out groups," all the team members together might look upon managers as the "out group." This is a rather typical phenomenon in organizations where an adversarial relationship exists between management and labor. TMM runs a risk in nurturing work teams in that the solidarity developed between team members could be used against management and management efforts toward goal achievement. If the work team norms and goals should become incongruent with Toyota philosophy, individual workers have been equipped with a powerful weapon to support counterproductive behavior and attitudes. The company team is the ameliorative to this possibility. Company team philosophy encourages team members to place their loyalty to the work team within the larger context of the cooperative unit that is the company team. That is why the work team alone is insufficient to realize organizational goals.

Before I address the consequences of work teams for team members themselves, it is appropriate at this point to return to the literature briefly since there is a close parallel between the work-team concept and the conclusions of Cole (1979) and Rohlen (1974). Both maintain that one feature differentiating the Japanese organization from organizations in the United States is the ability of the Japanese organization to utilize the primary group for organizational goal achievement. Cole, however, believes this is a culture-bound feature and cannot be adapted in the United States. He says, "Japanese employers do appear quite effective in the area of penetration and mobilization of primary work groups on behalf of organizational goals. This is an area, however, in which success rests most heavily on culturally unique patterns of behavior that cannot easily be absorbed by other nations" (1979, 251). Specifically he is proposing that the ability of the organization to penetrate and mobilize the primary group is a direct result of the unusual dependence—structurally, culturally, and psychologically—of the Japanese worker on the Japanese employing organization.

Toyota has not heeded Cole's cautionary note and has implemented organizational characteristics that attempt to utilize the primary group relationships of American workers for organizational goal achievement. As I have demonstrated, Toyota's efforts in this regard have been successful—at least initially and with some employees. It might be that over a longer period of time Cole's predictions will be affirmed. The dependence of Japanese workers on their employer is greater than the dependence of workers in the United States. Nonetheless, U.S. workers, maybe especially in this geographic region, are quite dependent on a "good" (in the sense of pay, benefits, and security) employing organization. This issue will be elaborated further in another section. It is important to note for our purposes now that the work-team concept follows a recognized pattern of Japanese organizational style and demonstrates many of the same consequences in TMM's case as the organizations studied by Cole and Rohlen in Japan.

Consequences for the Individual Worker. The preceding discussion of the consequences of work teams for organizational goal achievement presents a rather sinister picture of individual team members being controlled and manipulated to give 100 percent to Toyota, usually without their knowing they are being manipulated. To more fully explore this dimension, I will consider the consequences of work teams and what work teams mean to the individual team member.

When work teams are operating properly—that is, with members who are all cross trained on each others' jobs—job rotation is practiced. Since there are four or five jobs usually assigned to each team and the work day is broken down into roughly four equal segments, that means that a different task is performed in each major segment of the day. The benefits of this rotation schedule for the team member are obvious. First, it allows some variety of work, important in jobs that are intrinsically boring and monotonous. Interviewees recognize that production work requires the performance of small, repetitive jobs and appreciate efforts to insert variety and meaning, even if only on a minor scale. Second, team members are allowed—or required, depending on your perspective—to develop a variety of skills and learn several jobs. Through this, team members develop a larger picture of the production process and their role in it. This gives additional meaning and importance to what would otherwise appear as a relatively inconsequential contribution to the overall process of production. Third, job rotation can be a prophylactic for job-related injuries. If team members are rotating properly (i.e., rotating in such a way that jobs that could cause repetitive motion illness are alternated with ones that rest stressed body muscles), the likelihood of repetitive motion injuries is lessened. Additionally, the change in jobs decreases the potential of injuries caused by careless actions, the probability of which increases in direct proportion to the numbing monotony of the job. Finally, job rotation increases the feelings of camaraderie with other team members. Most team members seem to enjoy socializing with other team members. The fact that they all perform the same jobs contributes to closeness. Equally obvious is the fact that, for the same reasons, job rotation contributes to the overall goal achievement.

Job rotation is an ideal more fully realized in some departments than others. Wherever there has been high turnover of personnel, injuries, or jobs that have been redesigned, making previous cross training obsolete, the requirement to keep the line moving at the same pace may preclude job rotation. This will be the situation until team members can rebalance their knowledge of the team tasks. The situation can deteriorate to the point where the most skilled or strongest person becomes stuck on the most difficult team task without rotation, at least for awhile. Under these circumstances the chances of injury to the team member on the difficult job are increased and possibilities for rotation for the remaining members decreases even if they are all equally matched. Further, if the remaining members are not equally knowledgeable, rotation is likely to break down entirely. Not only does this pose an increased risk of injury to team members, but it can also demoralize the team. Team members were told in assimilation that rotation was the norm. If their actual situation does not conform to their expectations, frustration is likely to occur, in this way contributing to demoralization. Until this cycle can be broken with extra time or staff to allow cross training, rotation will not occur. It is for these reasons that I have consistently referred to cross training and job rotation as ideal conditions, not actualities present everywhere in the plant.

Another consequence of work teams, as they exist at TMM, is that they offer the opportunity for some intellectual challenge in what could otherwise be lobotomizing jobs. Especially relevant here is the opportunity and, indeed, the responsibility of team members to design ways to improve their jobs in terms of quality of product, cost reduction, and safety enhancement. The potential is that team members will feel that Toyota has hired them not only for their hands, backs, and legs, but also for their brains. With roughly 73 percent of team members with at least some postsecondary education, this aspect of the job could be critical in maintaining job satisfaction and preventing turnover.

There is some limit to how long job rotation and work improvements schemes will placate intelligent, ambitious, and well-educated team members. Another feature of teams, the low ratio of team leaders to team members, enhances the possibility of promotion. As mentioned previously, the team-leader position offers more intellectual challenge, variety, and prestige than the team-member role. The increased possibility of advancement and the positive feedback that advancement represents are benefits to team members that would be unavailable without work teams.

Work teams provide a warm social group experience for team members during work and outside of work. While some team members enjoy their teams at work but resent Toyota imposing on their free time to socialize in PT functions, others relish the opportunity to get to know team, group, and departmental members better and in a different context outside of work. They believe the social relationships benefit them personally and, as Pat Griffin, an engineer, adds, they contribute to smoother functioning at work. It is no small benefit to feel that there is someone on your own level you can turn to if you have a problem or need someone to talk to or someone you can feel close to. Certainly, all organizations provide an opportunity for their members to form this kind of close personal relationship with others. In the case of Toyota, even though the work-team concept does not deserve all the credit for what people do naturally, it does assist in primary group formation and nurturance.

Moving to a less obvious consequence of work teams is the change in perceptions and beliefs regarding the best way to get a job accomplished (i.e., team vs. individual) and how to solve human relations problems. Toyota teaches that team efforts are superior to individual efforts in goal achievement. The team members I talked to were true believers, even as some of them recognized that what they were getting was not what they were promised. Furthermore, some were true believers even as they described to me ideas that they, personally and individually, had developed in regard to job improvement. I also heard about many job improvements resulting from group efforts and the literature is replete with examples of team innovations. Nevertheless, the incongruity of singing the praises of team over individual accomplishments and, then, during the same conversation recounting individual achievements, did not seem apparent to team members. Perhaps, the newness of the work-team concept to team members and/or the special attention to that facet of the team members' experience encourages team members to selectively perceive team successes and ignore individual accomplishments. Whatever the reason, it represents an intriguing situation.

Team members' belief in team work is not limited to endeavors in production but has infiltrated other areas of their lives. Recall Lemming's argument for team dissertation research. Later in our interview he presented the case for team office work. Woulton uses the skills she has learned about group dynamics and how to conduct effective meetings in her activities at church. Gibbons describes the prepromotion classes on problem solving and job instruction as beneficial even if a person does not get promoted because of their general usefulness in a variety of areas of life.

In the reassessment of the consequences of work teams for individual workers, one must ask what is the effect of giving 100 percent effort to the job all the time. Several features of Toyota in addition to work teams encourage 100 percent commitment. These will be addressed in subsequent chapters. However, work teams should be viewed as a contributory factor—even though not a sufficient factor—in eliciting 100 percent work commitment from team members. I have demonstrated in the chapter on work context that giving 100 percent effort does not necessarily mean working 100 percent of the time. This point is not intended to diminish the effort necessary to maintain a 92 percent or 94 percent, or even a 78 percent pace. It is, however, an important point of clarification. It illustrates that 100 percent effort is not necessarily referring to the pace of work but to a commitment on the part of the team members to give 100 percent if called upon to do so. More than that, it demands that team members interpret flexibly what their duties are to Toyota. The commitment requirement goes beyond the narrow confines of "doing your best at your job" to "doing whatever is necessary" within, of course, certain moral and logistical bounds. So even though sweeping the floor or cleaning up a spill or "running" to get a tool to fix a piece of downed equipment may not be part of the job description, they are part of the job expectations for team members at TMM. At this juncture let us turn to exploring team member reactions to both the pace of work and the commitment required.

Interviewees had mixed reactions to this aspect of belonging to the team. Some describe being so tired after work that they literally had no energy left for family or other pursuits. This was a minority sentiment among team members. It was a more common complaint among group leaders, managers, and specialists. Most team members I interviewed maintained that, once they learned their jobs, the pace was steady and rather fast but not hectic. Within the ranks of team members, three elements were associated with attitudes toward the pace of work and the requisite commitment. First, attitudes varied by department. For example, interviewees from assembly, which is the largest department of team members, tended to have the most positive attitudes toward giving 100 percent. The importance of departmental assignment to attitudes is reinforced by team members, managers, and specialists as they consistently describe certain departments as high-morale areas and others as low-morale areas.

A second factor that influences the impression team members have of giving 100 percent to the job is their prior work experience. Few team members had prior manufacturing experience. This means they did not know what manufacturing work was like. They had no expectation of what would be asked of them. Neither did they know the tricks or the "bad habits" associated with work at some production facilities where workers are not required to give 100 percent. Therefore, they may be more likely to believe that Toyota's pace is the norm for manufacturing. Consequently team members as a whole may be less critical, more accepting of the pace than would workers with prior manufacturing experience.

Further, consider the former jobs that provide a point of reference for team member assessment of what Toyota expects of them. Fetalucci held five jobs at one time prior to joining TMM. Ruddlehouse worked at two so that she could support her family. Perkins is a college graduate who worked at a series of low-paying jobs while running a lawn-mowing service and building his own house. Dottard summed it up this way:

> Yeah, I have a good opinion of Toyota. Of course, if you would have worked at the type of job I had before. . . . I was paid salary. I was paid $17,000 to work as an assistant manager at (a fast food restaurant) when I quit. That's nights, weekends, holidays, and so on. And then all of a sudden, boom. Here comes this job that's Monday through Friday, pays $10.40 an hour, and I've gotten a few raises since then. Well, what can I say. They're my bread and butter.

All of the above team members feel giving 100 percent is a fair price to pay for what they get in return. Indeed, these are the team members who do not feel exhausted after work. For them, work at TMM demands less of their time than their previous job or jobs; work as a team member at Toyota is less invasive of other areas of life like weekends and holidays; work at TMM is left at work, after work they are mentally free; and the pay and benefits are the best they have ever experienced.

The importance of prior frame of reference is substantiated by comparing the attitudes of team members who did not have bad memories of former jobs. Gibbons and Lemming both fall in that category. Gibbons came to work for Toyota because although she liked her former job it did not pay enough, barely above minimum wage. She believes Toyota pushes workers too hard, as evidenced by the fact that "team members are not even allowed to use the rest room except at break time" and the fact that so many employees are injured. Lemming had to take a cut in pay to work for TMM. He called working for TMM "slave labor," but tempered that comment later, showing a common ambivalence that characterizes team members who express negative feelings, when he told me "They're being very fair, very open, very good people to work for, the Japanese are. They've been very fair to me although I did have to push them a little bit." Both of these team members work in a department with some of the worst work conditions in the plant and both have been injured on the job. Whether their attitudes result from prior frame of reference or their actual experience with Toyota is unclear. However, the prior frame of reference hypothesis is given further support by contrasting the team members with the managers and specialists.

Specialists and managers typically came to TMM from more lucrative positions (both in pay and job content) than the team members, and have more negative attitudes toward the commitment required by Toyota. Even though the jobs of managers and specialists do not force them to work on the line all day, 100 percent effort is manifested in other ways. These people generally work eleven to twelve hours a day and some time on weekends, for a reported norm of sixty to sixty-five hours per week. They do not get paid for overtime, except insofar as they are paid on a higher base than team members. They are expected to spend some of their "free time" in personal-touch activities. Since the work demands are different between team members and managers, it is impossible to know whether the work demands or the different point of reference cause the difference in attitudes toward 100 percent commitment. Both are probably important, nonetheless some support is added to the significance of reference point in attitudes toward expectations of 100 percent effort.

The third factor that influences how the pace of work affects team members is their physical conditioning in combination with their specific job or jobs. Obviously, team members who are in better physical condition and/or are not injured will not be as tired at the end of the day as those in poorer condition. Team members had to pass a physical examination before employment. However, this merely ensured that they had no major illness or handicap, not what kind of shape they were in. As team members gain experience in particular jobs, we would expect their bodies to adjust and strengthen. Accordingly, as the work becomes less physically exhausting, appraisals of the demands of work should become more positive. It makes sense that this relationship would hold if the normal performance of the job does not cause injury, if age and job monotony do not become important conditional factors, and if Toyota does not continuously adjust the takt time. (See Appendix C for a glossary of terms). I will make just a few comments about these contingencies.

It might be that the suggested inverse relationship between experience and tiredness on the job is more accurately a curvilinear relationship. That is, increased experience decreases the feelings of tiredness until a point where the age of a person begins to negate the effects of experience and tiredness again increases. I can only speculate on this issue from my research. However, several managers and a participant observer account from a similar facility in Japan (Kamata 1982) contend that the assembly operations at Japanese plants usually run at such a fast pace that only the young can keep up. As the workers age and need to slow down, they are moved to less physically demanding jobs or promoted to team-leader or group-leader positions.

If the pace of work is increased, team members will have to readjust to reach the same level of expertise they had before the increase. At start-up, takt time was relatively slow, more than sixty seconds. It was gradually decreased to sixty seconds and then to fifty-seven seconds. TMM maintains the alteration in takt time is necessary to allow more responsiveness to the market. In this way they contend they can avoid the practice of hiring and laying off workers with every fluctuation that occurs in sales. However, it may also be true that Toyota wants to take advantage of the newly acquired experience and anticipated better physical conditioning of its labor force by making greater demands on them. In any case, this will affect the relationship between tiredness and experience.

What can we conclude about the consequences for team members of giving 100 percent to work? The team members themselves whom I interviewed were generally positive, with negative attitudes associated with particular departments and with being injured. They were surprisingly positive as compared to other categories of employees, such as managers and specialists. Caution must be exercised in interpreting these conclusions since my sample was so small and nonrepresentative. Still it leads me to speculate that attitudes about work demands among employees can be partially understood by considering departmental assignment, prior frame of reference, physical condition of the employee, and changes in the pace of work, or takt time.

Work Teams: Problems

This section will elaborate work-team problems already alluded to within the text of the preceding discussion, suggest potential problem areas that have not been considered and provide a bridge between the work team and the company-team concepts. There are two categories of problem areas that are apparent from my analysis of work teams. The first category pertains to potential and actual problems that the organization encounters in attempting to form and maintain work teams. The second area relates to the actual and potential problems presented by the successful realization of work teams. In other words, the second category represents the problems that could occur if the organization is successful in overcoming the problems in the first category.

The major obstacles currently present within Toyota to work-team implementation and maintenance are failure of team leaders to follow the role prescribed for them in the Toyota philosophy, uncooperativeness among individual team members, and turnover among team members. These are issues identified by the interviewees themselves as areas that pose challenges to the work-team concept within certain departments of Toyota. I will examine each of these in turn before discussing the problems resulting from successful team formation.

The team-leader role is critical for successful realization of work teams. There is the tendency in at least some areas for team leaders to misunderstand their role or to choose to redefine it. Gibbons brought this to my attention as she described team leaders caught between the threefold pressures of meeting the daily quota, the demands of team members, and meeting quality standards. Juggling expectations from these three areas sometimes results in team leaders discouraging team members from pulling the andon cord and "encouraging" team members to work when injured. Without constant support and direction from upper management, team leaders may resort to resolving the dilemmas posed by these three demands in ways contrary to official Toyota policy and destructive to the work-team concept. Team leaders' innovative solutions to the pressure on them was particularly irksome to Gibbons because she felt they were manifestations of attitudes of managerial superiority. It is important to note that among the team members interviewed, Gibbons's was the most negative account of team leaders. Only one other team member mentioned a team leader who did not live up to Toyota standards. The frequency of this phenomena is not really important, however. The point is that there is a high probability of inadequate fulfillment of the team-leader role by at least some team leaders, and that this is a significant challenge to the actualization of the work-team concept. Team members working in teams with team leaders who do not live up to the Toyota model might be expected to feel that they are not getting what they were promised.

Despite all the screening, training, and manipulation of team norms, some team members will be uncooperative and even insubordinate. One of the features of work teams is cooperation between team members. If one or more members refuse to cooperate in the flexible give and take that is implied by this term, then the work team is in jeopardy. Lemming identifies this fear—unequal contributions from team members—as the reason Americans are reluctant to work in teams. Toyota attempts to prepare team members for this possibility and provide them with tools, both conceptual and real, for dealing with it. In other words, Toyota recognizes that this could present a major threat to work teams.

Team members interviewed had very high opinions of the members of their personal teams. Even so, some described incidents in other teams where a team member consistently refused to help other team members, refused to give 100 percent, consistently broke work rules (like tardiness or drinking problems that affected the work team) or were insubordinate to team leaders (Ruddlehouse, Perkins, Dottard, Axelrod, and Lemming). Two of these incidents were resolved with "corrective action" resulting in solution of the problem in one case and termination of employment in the other. The other two were not yet resolved at the time of my research.

The situation where individual team members decide it is not in their best interests to be cooperative because they are not receiving an equitable return for their contribution to organizational goal achievement poses a more significant threat to work teams. The above examples of team member uncooperativeness resulted from individual problems and/or idiosyncrasy. Uncooperativeness of the later kind results from organizational features and/or perceptions of organizational features, not uncooperative personalities. As such, when it is present, it is likely to be more broad based among team members and to have metastasized throughout various aspects of the organization. I saw no hint of this sentiment among team members I interviewed, even those in the skilled trades. The few comments I heard regarding perceptions of being personally exploited by Toyota came from employees who were not team members. Nevertheless, this should represent an important area of concern for TMM. One of the primary reasons for the necessity of the company team is to maintain the perception of equity among team members.

Under current circumstances, the phenomenon that is most problematic for work teams is turnover. Turnover, in the sense in which I am using it, refers to any change in the membership of those actually performing the tasks of the team. It does not necessarily mean that the team member has officially left the company, the department, the group, or even the team. It means the team member, for whatever reason, is not physically performing his or her normal role with the team. With this definition, turnover refers to vacancies created by team members who have been promoted, transferred, terminated, who have voluntarily left, or who are on restriction. Promoted, transferred, and terminated members are replaced. If the process does not occur too frequently, the team adjusts and goes on. Restricted members, however, have not left the team. They have been disabled in a work-related capacity and cannot perform their normal job yet still come to work and do "something." If the rate of restriction is relatively low and intermittent, team leaders can fill in for restricted team members with the restricted team member performing some of the duties of the team leader such as answering the andon cord and restocking supplies. When restriction become chronic and pervasive throughout a department or a group, and especially if combined with other forms of turnover, there are simply not enough team leaders to fill all the vacancies. Team members must carry an extra burden. Healthy work teams must be broken up and rebalanced with those whose ranks have been depleted of experienced team members. Assimilation and training for new members must be shortened, and temporary employees or team members borrowed from other groups must be utilized. In addition, this becomes a vicious circle as job rotation and cross training are attenuated, becoming a luxury that the group can no longer afford. Healthy members transfer out as soon as they acquire enough seniority and the voluntary separations increase. Work teams deteriorate under these circumstances due to demoralization of members and reassignment of tasks from teams to individuals in order to meet the line demands of the day.

On restriction will be elaborated in more detail later. It is sufficient here to know that the injury rate at TMM during the first couple of years of operations has been high by industry standards and that the results of this have been hardship and suffering for the injured team members and the remaining healthy team members as well as a serious threat to the work-team philosophy in departments with the highest rate of incidents.

Although there are problems in work-team formation and implementation, by all accounts TMM has still been fairly successful in their development. But even in success, work teams pose problems. I have already examined these in some detail in the section on consequences of work teams. Briefly, they are the possibility of work teams developing into ingroups, with the resulting intraorganizational rivalry and conflict that this could create, and the effect on team members of giving 100 percent effort to their job. As noted in the prior discussion, giving 100 percent effort is not generally perceived by interviewed team members as detrimental yet, but varies by department and may change as team members age and gain experience, as well as with adjustments in the pace of work. I bring these problems up again at this juncture in order to highlight the necessity for linkage between work team and company team. To elaborate this notion, I present the Mazda Company philosophy as communicated to workers at their new facility in Flat Rock, Michigan, during their orientation to the company:

> This message was driven home by one orientation speaker after another. Mazda cares about you, not just as a worker but as a whole person. Your safety was our number one priority when we designed this plant: we chose the safest tools and equipment we could find, and we tilted the overhead assembly line, so you could work under the bottoms of cars without craning your neck the way the workers at GM and Ford do. We built this giant fitness center, where you can play sports or exercise. It even has a driving range. At American companies only executives get perks like this, but here we're all part of the same team. We're going to offer you classes in Japanese language and culture and other non-work subjects to broaden your horizons. We're going to cross train you in advanced skills like robotics, so you can rotate jobs and have a more varied and rewarding work life. If you're having problems at home—your kids are sick or having trouble at school—tell us, and we'll do what we can to help arrange your work schedule around the problem. The emphasis here is on pulling together like a family. (Fucini and Fucini 1990, 71)

According to Fucini and Fucini, the philosophy was only rhetoric. The heightened expectations it engendered in workers led to disillusionment and cynicism when the philosophy was not actualized. Fucini and Fucini comment:

> But it was not only inexperienced workers (who accounted for over 70 percent of the plant's work force) who were disillusioned at Flat Rock. Those team members who had worked at American auto plants also were dismayed by Mazda's JIT work pace. More jaded than their younger teammates, the Big Three veterans had been skeptical about

Mazda's orientation pledge to involve workers in the consensus decision-making process, so when the team system
collapsed, they had taken it more in stride. (1990, 149)

The Fucini and Fucini interpretation of the Mazda situation demonstrates that without the articulation of the company team in the day-to-
day interaction of employees, the problems considered above have a
high probability of occurring. That is to say, an organization that
attempts to implement work teams without corresponding implementation of the company team will not only fail to achieve the intended purposes of work teams but will, in all likelihood, create a force that is
counterproductive to those purposes.

The Company Team

Company team is the second major conceptual category in the theory I
constructed to explain what work is like at TMM. Like work team, company team is a theoretical construct and describes an intended state, an
ideal type. We can see that if Fucini and Fucini are correct in their
assessment of Mazda at Flat Rock, reality can be quite different from
intended states.

Members of the company team share a common goal and work
cooperatively to achieve that goal. There are feelings of equity and reciprocity among members. Even though only production and maintenance workers are specifically entitled team members, all TMM
employees are team members of the company team. When the senior
vice president tells employees in assimilation, "We're all members of
the same team here," he is referring to company-team membership.
When the plant manager quoted in *Toyota Today* commented about the
hiring process by saying, "By the time candidates get through the process, they realize it's not just a single job they're applying for, but membership in a team," it is company-team membership he is talking about.
When production worker Perkins said, "The management is very open
to suggestion. Everybody is a team member trying to achieve a common
goal," he meant that all employees are members of the company team.

The essence of company team implied from the above—as well
as and many other similar comments in publications, interviews, and
informal conversations with Toyota employees—is that all employees

of the organization share a common fate, work for the same goal, and put aside differences that arise between individuals and groups of employees (e.g., management and labor, office staff and line workers, electricians and tool and die makers) for the benefit of all. An effective company team pulls together the often diverse orientations and goals of individual members, departments of the organization, and the myriad professional categories present into a coordinated effort toward organizational goal achievement. The company team is the shared belief in the community of fate and a shared set of attitudes toward the organization, work, and other employees based on the community of fate.

Unlike the work team, the functioning of which employees see daily, the company team is an abstraction. A person can walk throughout the plant area while Camrys are being produced and see what most work teams do, but she or he cannot see the understandings, agreements, and relationships that unite all those employees into the Toyota company team. Since the company team cannot be physically observed or experienced, employees have to rely on indirect indicators to provide evidence of its existence. My elaboration of the company team at TMM examines what I think are the more important indicators: policies and practices that promote feelings of equity among employees (abbreviated henceforth to "feelings of equity"); open communications; strong human resources department; bureaucratic flexibility; job security; and individual importance.

Feelings of Equity

To understand the notion of equity at TMM we need to explore the kind of employment agreement interviewees and informants believe they have with Toyota. A typical employment contract at a Western organization, especially for nonprofessional employees, specifies an exact amount of time rendered in the performance of specific tasks by the employee or an exact amount of service or products in return for a prearranged wage. In contrast, the Toyota agreement with its employees has two parts. The first part is like the normal work agreement just described. It specifies minimums—minimum number of hours worked by employees for minimum pay from Toyota. The second part, the tacit agreement, contains the expectation by Toyota that employees will give 100 percent in commitment and effort; that they will contribute ideas to improve the quality, effectiveness, and safety of production; and that they will flexibly define their jobs so that they do whatever is necessary to get the job done well. In return, employees expect to receive more than the minimum wages and benefits outlined in the contract. They anticipate an equitable share of the fruits of the organizational success achieved through their extra efforts, including a share of the psychological rewards of success—like recognition, feelings of accomplishment, and pride—as well as monetary bonuses.

This type of agreement fosters feelings of ownership among employees. Team members identify with the product and Toyota. My informants used the pronoun "we" when referring to Toyota, and "our Camry" when discussing the Camry. I think this is especially remarkable since the product was developed and produced in Japan before it was made at TMM. The assistant staff were the least likely to identify with the product and more likely to use "they" and "us" frames of reference. In interviews with assistant staff personnel, the Camry was never mentioned. This was not the case with interviewees in other positions.

Employees have to trust that Toyota will carry out its part of the tacit agreement because they must make their contribution first. Then, when and if that contribution results in added value, Toyota is expected to share the fruits of everyone's labor equitably. The issues in this equation that require trusting Toyota are the definition of added value and equitable distribution. TMM has certain organizational structures and policies that encourage employees to trust that these decisions will be fair to them. These features send the message that Toyota will carry out its part of the bargain, and in so doing, they nurture belief in the community of fate.

Employee Selection and Assimilation. Nurturing the community-of-fate belief begins in employee selection and assimilation. Recall that the majority of employees hired had no manufacturing and no union experience in their background and tended to be young and well educated. Without prior manufacturing and union experience, employees are less likely to have been in a comparable situation where an employer violated employee trust, or where the accepted belief was that the employer would violate trust. Therefore, they are less likely to be cynical of the community of fate. The difference in the reaction of experienced manufacturing workers compared to inexperienced workers is reinforced by the Fucini and Fucini study of Mazda. As pointed out earlier, the Big Three veterans were cynical of Mazda claims from the start. They viewed other workers as naive for not seeing the company philosophy as a mechanism for wringing out as much effort as possible from workers with nothing extra in return from Mazda.

The selection and indoctrination of American managers is perhaps even more important for the promulgation of belief in community of fate than team member selection. If the American managers, who have more contact with team members than the Japanese managers, do not believe in the system and do not carry out, on a day-to-day basis, the spirit and letter of the agreement, than team members will begin to doubt Toyota's sincerity. High-level American managers hired by Toyota did have previous manufacturing experience, some with competitors of Toyota. The potential is greater that they have brought with them entrenched values about the roles of management and labor that are contrary to Toyota's philosophy on the matter. On the other hand, since so few of them were hired in relation to the numbers of team members, more care and time could be spent in selecting the right people. Furthermore, not content to hope that the selection process had weeded out managers who might be destructive to the philosophy, Toyota paid considerable attention to the training and monitoring of managers. The critical nature of these positions is recognized by Toyota in that each American manager is supplied with a Japanese coordinator and/or supervisor to teach him or her Toyota's method of operation and philosophy, to assist him or her in a variety of ways and to facilitate communication with other Japanese staff and the home office in Japan. In this way, Toyota has defused the power and impact of American managers at least until they have proven their understanding and commitment to Toyota procedures and philosophy.

Many lower-level managers in the production areas, such as group leaders, did not have prior manufacturing experience. They come to the position of group leader with backgrounds in government civil service, fast-food restaurant management, teaching, and so on. Like the inexperienced team members, these group leaders should be more malleable, more accepting of Toyota's definition of their role in the organization than would managers with manufacturing experience. The preferred policy now is to promote from within the ranks, so that in the future, managers will be fully conversant with Toyota philosophy, and only with Toyota philosophy.

Assimiliation represents the first formal introduction of the team concept (work team and company team) to employees. Perkins described to me his introduction to the company-team concept.

> Everybody is a team member here. That's what we were told when I was having my assimilation training. The American senior vice president came in and said, "We have to have the titles, management, and so forth, to know what our duties are. But basically everybody is a team member trying to achieve a common goal. We build quality into every product we build. Whether it be in the office, in the assembly, or wherever." After he said that, I said, "Yeah, right." Yet I see it. The management will try to make as many contributions, as far as advice, as they can. They want to hear from us what we can do to better our work place and the quality of our product. I'm sold on them.

The Leveling of Hierarchical Distinctions. Perkins elaborated on one factor that convinced him that the vice president's words were not just propaganda: "There's no real differentiation between team members to the extent we all wear the same clothes. You don't see them walking around in a suit and tie. You don't call them Mr., you call them by their first name."

Toyota supplies each member with uniforms and encourages all members to wear them, although it is not required. The uniforms are consistently worn by management and professionals, with team members in production and assistant staff most likely to deviate from form. Pat Griffin, a specialist, explained why he always wears the uniform:

> I wouldn't have been one of them, this way I'm one of them. I personally like a manager who is a working manager, who rolls up his sleeves and works right along with the people. I don't like managers who are there just for PR. We don't need them. You feel less intimidated by people and you feel more politically able to give your opinion on something if you're all wearing the same clothes.

To expand on this theme, other usual management perks are absent from TMM. Everyone contends equally for a parking slot close to the plant. There are no reserved parking areas for managers. Several cafeterias exist but are located to facilitate easy access for employees assigned to various areas in the plant. Managers, specialist, and team members use the one closest to their work area; there is no separate dining facility for managers. Lemming told me how the departmental managers often ate lunch with him and his team members. He was quite impressed because on one occasion Mr. Cho, then president of TMM (who happened to be the only TMM manager or executive consistently addressed by all American employees as Mr.), brought his lunch tray to the table and ate with Lemming's team, carrying on a relaxed conversation with them. Managers and professionals seldom leave the facilities for lunch. This is so because they usually do not have time, but also they are encouraged to interact with other employees at lunch and to not exercise prerogatives unavailable to team members.

True to the form present in Japan, TMM has an open office format. None of the TMM executives have private offices. All have desks in a common office with other members of their department. There is nothing to distinguish the president's or other executives' desks from any others, except for the slippers located under the desk of one Japanese executive. In some office areas the largest, most impressive, most personalized desks belong to the clerical staff. This feature makes executives appear more equal to other employees, makes them appear more accessible and indeed, promotes a casual, open-door policy, since there is no door to shut. Meetings that require privacy can be held in one of several meeting rooms available for that purpose. However, the majority of office business is attended to at the desk.

The physical arrangement of the office is most frustrating to American assistant staff members. The assistant staff members I talked to found it difficult to concentrate in the midst of all the "chaos and distractions." Every phone in the room, as well as conversations across the room, can be heard at almost every desk. People face each other across their respective desks with another desk to their back and side. Most offices are crowded. In some offices there is no natural path for traffic because desks are arranged in almost a random fashion. When I asked assistant staff what organizational features they would like to see changed, they overwhelmingly said the open office setting.

In most American organizations, management perks include flexibility in scheduling time. This is not the case at Toyota, at least not officially. Managers are expected to be present whenever the production team members they support or supervise are present. Griffin told me how extra effort and willingness to do whatever is necessary was rewarded by his previous employer: "Our management was very good about 'Hey, you worked all night. Friday afternoon, you got any meetings? No. Then I don't want to see you here after 12:00. Go home.'"

The stipulation that managers and specialists must be present whenever the production line of their shift is running prohibits this kind of flexibility and rules out using flexibility as a reward to managers and specialists for "busting your butt for Toyota." In combination with long hours, this equates to "You have to give up seeing the doctor, the dentist, getting your driver's license, coaching little league baseball, or attending a PTA meeting" (Eric Klaske). While this is an exaggeration—management staff can take off for doctor's and dentist appointments—it indicates how frustrated some managers feel about the policy. This is the only management perk that the managers and specialists I interviewed resented losing. Managers and specialists expressed the sentiment that, while this policy might contribute to the feeling of equality among all company-team members, it does so at the expense of equity for managers and specialists.

Bonus System. The policies that level the privileges of management to conform more closely to the work situation of work-team members are accompanied by more tangible disbursements of the added value contributed by employees to organizational goal achievement. This takes the form of a benefit package—which has gradually become more generous since start-up—and a bonus system—called "performance awards." Twice yearly, TMM may provide performance awards to all employees, dependent upon management's assessment of goal achievement. I do not know how this decision is made, but it certainly involves more than consideration of whether or not a profit was made at TMM. The corporate management seems to be as concerned with market share and quality ratings as it is with profit. If TMM were treated by its corporate team like a local profit center, the cost of building the new facility would preclude profits—and thus bonuses—for several years. Yet, performance awards were distributed in the second year of operation and consistently thereafter. It is in this respect, that corporate team membership facilitates implementing aspects of the company team. Without the resources of the corporate team, bonuses at this early stage of operation would almost certainly be impossible.

Individual performance awards are equal to a percentage of each person's wage. Since within the categories of production team members, skilled team members, and team leaders the wages are essentially the same, all team members in the same category receive the same performance award regardless of their individual performance. Equal performance awards maintain the principle of equal pay among work-team members. Further, the concept of equal awards communicates to employees that it is overall company-team performance that matters and that individual performance is important only insofar as it contributes to company-team performance. Each individual employee can increase the amount of their personal reward only by helping the company to be more successful. Initially team-member categories received higher percentage awards than some of the management and specialist categories. The performance award system for specialists and managers has since been adjusted by raising their percentage level and introducing the element of merit into the determination of individual awards for specialists and managers.

Past performance awards were distributed just before Christmas and in early June just before most employees take vacation. The timing of the awards is mentioned by employees as an indication of the thoughtfulness of TMM. The timing creates the impression that one award is a Christmas present from Toyota to the employee and the other is monetary assistance so that employees can have a more enjoyable vacation. The facts that performance awards are entirely at Toyota's discretion and that few employees are sure exactly what factors are considered in the decision mystify the process and reinforce the belief that the award is a gift, a demonstration of Toyota's generosity, not a right due the employee.

The contribution that performance awards make to the belief among employees that Toyota will honor its part of the tacit agreement (referred to above) is enormous. Many employees know that TMM could not possibly be showing a large profit, even though performance has been good by other standards. To ensure that all employees are aware of Toyota's generosity, on at least one occasion Mr. Cho has sent letters to each explicitly reminding them to put wage and benefit improvements into the context of the relative youth of TMM and the recessionary situation at that time in the industry. The intent is clearly to impress upon employees that Toyota is not only honoring its part of the agreement but is willing to trust in the future performance of employees.

Nonmonetary Awards. Important as monetary rewards are, sharing the benefits of successful goal achievement involves more. In order to encourage belief in the community of fate, TMM has established several procedures, the purpose of which is to distribute the psychological benefits of organizational goal achievement. These procedures attempt to provide recognition through praise and other forms of positive feedback to team members for a job well done. For example, when a department or group is able to maintain the 100 percent pace for a shift, they are recognized throughout the plant by an announcement, a small gift commemorating the event, and possibly coverage in Toyota publications. The first time Toyota was recognized nationally for the quality of the Camry, employees were greeted at the entrances to the facility with banners praising their accomplishment. In addition, Mr. Cho sent a letter to each employee thanking the employee for his or her personal contribution and Toyota bought each employee lunch and a t-shirt in honor of the occasion. As you can see, the feedback often takes the form of inexpensive tokens like t-shirts, placards, or pictures and are almost always accompanied by ceremonies or company announcements and recognition in a Toyota publication. The relatively low cost of the items is not important in the scheme of creating an atmosphere that supports belief in the company team. To the employees I interviewed, they symbolize recognition and the sharing of the feeling of accomplishment that comes with goal achievement. Here again we see a distinction between team members and other categories of workers. The frustration caused by a lack of feelings of accomplishment was a pervasive theme for lower-level managers and specialists.

Powerful Human Resources Department. Another organizational feature intended to contribute to feelings of equity among company-team members is the role that the department of human resources plays in the organization. *Toyota Topics* (June 1990) highlighted the role of the human resources department in a special section, and it was referred to often by interviewees. Human resource staff are represented on the panels that interview all employees hired by TMM. There is a human resource representative (HR rep) assigned to each department so that all employees will have someone in human resources who is familiar with his or her work area and is accessible whenever questions or concerns occur. The presence of an HR rep is required before any disciplinary action can be taken against an employee and his or her input is required on every promotion decision pertaining to members of the lower hierarchical levels. Ostensibly, the role of human resources in each of these situations is to serve as an impartial advisor and/or mediator and to ensure that the company rules are administered in a consistent, fair manner.

Two managers (Huntington and Slider) who had worked previously at other manufacturing facilities remarked about the exceptional power of human resources at Toyota compared to their former employers. The Japanese managers interviewed indicated that a strong human resources department is an important characteristic of Japanese management implemented at TMM. Huntington explained that a powerful role for human resources is necessary to ensure uniform adherence to rules and policies throughout TMM and across all levels of the hierarchy. He believes that inconsistent implementation of policies and rules contributes to charges of favoritism and feelings of injustice and unequal treatment among company-team members. Thus, he argued, only by enhancing the power of human resources over the various other departments within TMM can consistent and equitable treatment of employees be possible. This is the logical support for the power of human resources and demonstrates how vesting power in human resources should contribute to the company team. However, not all members of management share Huntington's conviction.

Some specialists and managers complained of the power of human resources in that it dictated the inflexibility in their work schedules and contributed to coddling the team members (Slider, Griffin, and Klaske). Twice members of this group referred to the "gutless" character of human resources demonstrated by the fact that it is afraid to incur the displeasure of team members by giving "equitable" rewards to managers and specialists (in terms of time scheduling, overtime pay, and bonus pay) and in usually siding with team members whenever there is a dispute between the groups (Griffin, Klaske). Prichard described as a "dirty little secret" the pressure that group leaders feel to please team members. He maintained that the group leader is held accountable by human resources for any negative attitudes expressed by his or her group members in opinion surveys regularly conducted by human resources. In essence this gives team members a high degree of power over the group leader and makes group leaders feel as though they are caught between powerful team members and demanding upper-management levels.

If indeed human resources is successful, even minimally, in assuring that employees at all ranks are treated equally, those who stand to lose the most in the process would be expected to be the most resentful. Certainly managers and professionals as a category have lost the most at TMM in comparison to the favored status they would enjoy in many organizations. Therefore, the expression of some resentment regarding this issue, primarily by the lower-level managers and specialists interviewed, affirms the power of human resources relative to other departments, affirms that this power is used on behalf of the interests of team members, and also affirms the hazards inherent in defining exactly what equity means in a way that satisfies all company-team members. The team members interviewed did not seem to be aware that they were being coddled by human resources, nor did they express knowledge of any advantage to them resulting from having such a powerful advocate.

Slider likened the HR rep's role to that of the union steward in unionized organizations in the sense that the HR rep is an advocate, an ombudsman of the hourly workers. At minimum, one might speculate that placing HR reps in roles frequently played by union stewards is an attempt on the part of Toyota to co-opt the union steward position within its structure in order to forestall the attraction that unionization might present to the rank-and-file workers. In fact, taking this perspective lends a new dimension to our understanding of the power of human resources within Toyota. It was suggested by Griffin and Klaske that the reason human resources "coddles" team members is to prevent any dissatisfaction among team members that might lead to their interest in unionization. In other words, Toyota's perception of the threat of unionization has given team members more power within the organization than they might otherwise have. This power could explain why Toyota is more concerned with equity for team members than lower-level managers, if indeed that is so, and why a powerful human resources was necessary to act as a powerful advocate for team members in relation to management.

Open Communications

There are three important components of open communication as it is used to promote company team at TMM. The first element is that certain information available to management is communicated to workers at all levels of the hierarchy. This information can take the form of the rationale for certain decisions, the history of the organization, and information about environmental factors—especially the market, competitors, suppliers, and government. At TMM, information sharing of this form begins in assimilation and continues through a variety of meetings, memos, letters, the open-office format, the accessibility of managers and HR reps, prepromotion classes, groups formed for various purposes whose membership cuts across organizational levels (such as quality control circles, the United Way board, and the team member association), a closed-circuit TV network that broadcasts to monitors located throughout the plant, and Toyota publications.

The second component is communication upward from lower-level workers to upper management. Some of the features mentioned above as providing a format for downward communication also allow upward communication. Any characteristic that presents the regular possibility of informal conversation between management and workers is a conduit for both upward and downward communication. For example, any manager who frequently sits with team members at lunch, attends parties with team members, and sits on a United Way board with team members is going to hear what the team members think, are concerned about, and have gripes about, even if in a somewhat guarded fashion. At TMM there are, in addition, specific organizational mechanisms that facilitate the movement of communication upward. These are regular opinion surveys and the hotline. Results of opinion surveys are reported to departmental managers who communicate them to lower-level managers. The expected procedure is then for the results to be discussed in a meeting of departmental managers, human resource personnel, and team members.

The hotline allows employees to ask a question or make a comment anonymously. Hotline queries are directed to the human resource department, which then finds the answer to the query or comment and posts the query and whatever response is appropriate on a hotline bulletin board accessible to all employees. According to a special section discussing the hotline in *Toyota Topics* (Nov. 1989), human resource personnel do not answer most of the questions themselves. Instead, they direct the query to the appropriate department or position for a response. Several interviewees mentioned the hotline and some of the concerns expressed. Dottard described complaints aired through the hotline about the choice of radio stations broadcast in the production area and the temperature in the facility during one summer. Other hotline complaints concerned what team members saw as inadequate implementation of Toyota philosophy or inconsistent application of some rules.

The third and perhaps most crucial element of open communications is the attitude among employees that the information they receive is honest and unbiased and that their communication to upper management will be attended to. That this attitude exists to some degree at Toyota is illustrated by one team member who said,

> They tell you what they're going to do and they do it.
> There's nothing hidden there. They don't go behind our
> backs and do things. When we have departmental meetings,
> team members will go ahead and ask questions—whatever
> kind of questions—at the end of the meeting. They are
> always answered. They're real committed to the employees
> here. (Perkins)

Another example is provided by Martinez, an electrician who relayed the standard procedure for reporting maintenance problems. He said that any time a problem is encountered—for example, a machine is not working correctly—band-aid repairs are utilized to keep the line going. When there is a safety hazard with a piece of equipment, the team member involved makes a sketch of what broke, what went wrong.

> Underneath that you put what you did temporarily just to get
> it running, and at the very bottom any suggestions regarding
> reoccurrence prevention. You report this to the team leader
> and the next shift coming on. All these reports have to be
> signed off by so many. They go all the way up to engineers

and the plant manager. That's nice too. They know about and take an interest in the daily things that go on.

According to Fetalucci, team member ideas are given serious consideration:

> Here they look at your ideas. You have a lot of people looking at your ideas, from the group leader up. Everybody has to sign that form. I think even the plant manager signs that. You put an idea into the suggestion process, it's considered. And you feel good that your ideas are considered.

To provide a contrast between the attitudes displayed in these statements of company-team members and the situation at a facility without open communication, consider Prichard's comments about his new employer: "Here managers don't manage, supervisors don't supervise and workers don't work." When asked to explain what he meant by managers don't manage, he said,

> I've been there almost four months now and I don't think my supervisor knew my name the first two weeks I was there, hardly spoke to me the first two weeks I was there. I don't think I've ever even shaken hands with anyone above the level of foreman. I've never seen a member of management out on the floor except running around on little go carts. I see no evidence of management reaching out to staff to try to communicate. It's basically an adversarial situation—"We're the bosses. You're the worker. Here's the union contract. You have to do this."

Open communications affects the implementation of the company-team concept in two respects. First, through the process of communicating information regarding the history of Toyota and Toyota operations and systems, employees gain a sense of what Toyota is and the part their contribution plays in the total scheme of Toyota operations. This facilitates identification with Toyota and belief in the community-of-fate ideology. For example, interviewees who had not been to Japan all had stories about how things were done in Japan and how that affected

TMM operations. Fetalucci told that in the Japanese operations things are never written in ink. The reason for this, according to the story, is to allow for rapid change and flexibility. This story was communicated to him by his group leader and interpreted by him as a demonstration of Toyota's commitment to change and improvement. According to Woulton, labor-management strife in the early history of Toyota in Japan has led Toyota to the current philosophy—adopted in the United States—of treating workers humanely. Even though Dottard expresses consternation regarding the necessity of going through maintenance and engineering to get some kaizens implemented, he also recognized the reasons for involving these departments. He told me that only these departments can see the larger picture of the impact a kaizen in one department might have on other sections and only they can balance the production process after a change has been made in one area. Perkins described the principles of the kanban system of inventory control and the advantages of just-in-time manufacturing. I could go on with examples, but the point is that all interviewees involved in production and maintenance functions demonstrated repeatedly an understanding of the history of the company, the total production process, and the contributions made by various sections and aspects (i.e., the andon cord and board, kanban, and just in time) to the total organization.

Second, all employees are informed on a continuous basis about the market for the Camry, about the activities of other automobile manufacturers, and about any environmental factors that might impinge on Toyota operations (such as government regulations and public opinion). By raising the consciousness of employees regarding Toyota competitors, feelings of "we are all in this together" are nurtured in the same way that esprit de corp is created within children's groups in summer camp by designing competitive activities between groups.

Television monitors located in the break areas keep team members informed about the product market and competitors. Additionally, the content of monthly meetings in some departments is partially devoted to market news and *Toyota Topics* always contains details regarding the market as well as relevant government and competitor activities. By raising the specter of outside competitors as a major organizational concern and attempting to include all company-team members in the rivalry between Toyota and its competitors, Toyota encourages all team members to identify with Toyota and believe in the community-of-fate ideology. It should also be noted that during this time period many American competitors were closing plants and laying off workers. Obviously, through dissemination of this kind of information, workers are forced to compare any gripes and discontents they might have with the prospect of layoffs that faces similar workers at competing auto manufacturers.

The technique of increasing internal group cohesiveness by emphasizing a common external threat, or rival, is common in Toyota's Japanese facilities according to Dave Ohiu, a Japanese Toyota executive (Japanese trainers and executives often assume an Anglicized name in interaction with Americans). He described competition as a very important motivator of workers in Japan. He believes that the Japanese workers respond more to competition as a motivator than do American workers, at least for the present. In light of the company-team concept as it has been developed here, I believe we can see this not so much as differential response to competition as a motivator but, rather, as differential experience with the prerequisite trust that underlies the community of fate. Without acceptance of the community-of-fate notion that all employees have an equal stake in Toyota's success, we would not expect workers to identify with the organization against its competitors.

Bureaucratic Flexibility

Instances relating to and feelings regarding bureaucratic flexibility and the lean work force rationale were described by almost every person interviewed. Bureaucratic flexibility is not an oxymoron in the sense that it is referred to by team members. Fetalucci called it "working with the policy." By this he meant that Toyota bends the rules for the benefit of individual employees. This is not just a passive turning a blind eye to rule violations but an active effort by Toyota to make the system accommodate individual needs. Almost every employee had a story illustrating bureaucratic flexibility. Woulton told of a team leader who voluntarily transferred to second shift. After a period of time, he found that second-shift work was disrupting his family life and causing other personal problems. He petitioned to be able to return to first shift as a team leader, stating that, as the situation currently existed, he would have to choose between the welfare of his family and keeping his job with Toyota. Under those circumstances, the team leader said he would have to quit his job. The policy is to allow transfers to first shift on the basis of date of request, with those who ask for transfers first getting them first. There is a long waiting list of team leaders desiring to transfer to first shift and Woulton's acquaintance was close to the bottom of the list. Nonetheless, Toyota bent the rule in his case and allowed him to return to first shift.

Dottard and one other team member in Dottard's group were each very interested in being promoted to team leader. They both had taken the prepromotion classes but had not yet passed the last test that is given some time after completion of the class. In the meantime, an opening for team leader had occurred in their group. Rather than pick a team leader from among those already eligible, the group leader and HR rep decided to wait to make a choice until Dottard and his teammate could complete the testing. Dottard said,

> It would have been pretty easy for them to say, "Well, you're not really through the classes yet. Catch it next time." And there wouldn't have been much we could say. I would have understood that. But they actually held it up for us. They went out of their way for us to have a chance at the promotion.

An example provided by Perkins is about a team leader whose wife wanted to meet him for two weeks of vacation immediately after his training in Japan. The only way they would be able to do this was if Toyota were to schedule the team leader's return flight two weeks after the rest of the group returned and allow him to take his vacation in conjunction with his training in Japan. The vacation timing required no special consideration, but the flight scheduling did. Since Toyota was willing to do this for the team leader, even though it required a rather small accommodation by Toyota, it demonstrated to Perkins that Toyota was concerned about employees as persons.

Ruddlehouse described a situation with a team member in her area who had a drinking problem. She said Toyota could have fired him. He repeatedly violated the rules regarding attendance and performance. Instead, his supervisor and the HR rep worked with him, forced him to join AA and see a counselor. In Ruddlehouse's words, "Toyota really stepped on him and helped him straighten up his act. That helped me respect them. They really are here to try to help people. I think there are a lot of ways where they really do have the people in mind."

Several interviewees told me about a particular team member who was injured on the job. Lemming gave this account of the story: "Toyota could have kicked him out the door and there is nothing he could have done about it. Instead they retrained him and now he is working in another area." Clearly the importance of bureaucratic flexibility as represented in these accounts is beyond the direct impact it had on the individuals involved. The stories generated by these incidents are told and retold among employees creating and reinforcing an impression of the significance of the individual to the organization as well as the nature of the relationship between management and labor at Toyota. Management is viewed as "more reasonable" by Prichard, the former production group leader, than management at most American facilities. He clarified this by saying that Toyota management is more flexible, more open to new ideas than management at other places he has worked. Griffin echoed this theme when I asked him why he no longer belonged to the professional engineering associations to which he formerly belonged. He said that since working for Toyota he has not had time for professional activities and also that Toyota does not encourage affiliation with professional societies in the same way that his prior employers did. Then he added,

> However if somebody went to them and said, "Hey, how come we don't have a way to belong to our professional associations?" I'm sure they'd say, "We'll look into it and we'll figure out a way." And then in a couple of months or a year, they'd work something out.

There are two attitudes expressed by Griffin's statement; the belief that management is reasonable and, expanding on that assumption, a willingness to give Toyota the benefit of the doubt. Again, this refers to the element of trust mentioned previously. Griffin trusts that Toyota, if made aware of the problem and if given enough time, will correct the oversight, mistake, or injustice. This implies an extension of the tacit agreement referred to previously that if Toyota is reasonable, flexible, and humane in its treatment of employees, the employees will reciprocate by viewing Toyota mistakes as benign oversights that will be rectified when Toyota is made aware of the problem.

Bureaucratic flexibility has a downside as well. There is a high probability that what is perceived as flexibility by one person and within one context will be viewed as favoritism or capriciousness by a different person or under different circumstances. Any time exceptions are made to rules for a particular individual, the rules are being applied unevenly and inconsistently. So, for example, the other team leaders waiting to be transferred to first shift might have resented that the team leader in the Woulton story was moved to the top of the waiting list and transferred ahead of them. In this case, bending the rule benefited one person at the expense of others. Whoever has the power to apply the rules flexibly can use that power to make the system more humane and responsive to individual needs or that person can use the power to further personal ends or assist personal friends. Whether company-team members interpret instances of bureaucratic flexibility as the company accommodating employee needs or as favoritism and arbitrariness depends in part on the rather fragile, amorphous trust. Also, it depends on how management in fact implements this policy. Thus, it is imperative for the maintenance of the trust atmosphere that the power given to managers to make exceptions to rules is not abused by them. The fact that personnel from the human resources department take part in most of these decisions underscores their pivotal role in Toyota's strategy to create the company team.

Human resource personnel should provide a check to the possibility of managerial favoritism and arbitrariness. Even if their power is limited in this respect, the policy of including human resource personnel helps to create the impression among team members that management can violate the rules only when it is in the best interests of the company team to do so.

Job Security

A long-term commitment between Toyota and its employees is an absolutely critical ingredient of the community-of-fate ideology and its actualization. Feelings of equity and bureaucratic flexibility depend on a long-term relationship, and open communication is enhanced by it. Every interviewee mentioned job security in one form or another as either the reason they took a job with Toyota and/or the reason they would remain with Toyota even if offered a more financially lucrative position with an American competitor. Without a doubt, job security represented to my sample the most important advantage of working for a Japanese organization.

There are no written guarantees of lifetime employment for American employees of Toyota. Even though unwritten, employees clearly expect that unless they "screw up royally," they have a job for life. The lifetime employment commitment was premised on the beliefs that Toyota is well managed and has a good chance of future success, that future success will be directly influenced by the individual employee's commitment to the company's goals, and that Toyota will not cast people aside when there is a downturn in the market or when the individual is less useful to the organization. The emphasis in this orientation is on the proven track record of success of Toyota and on the contribution of individual dedication and hard work to Toyota's future success.

These beliefs allow employees to place the locus of control for job security with themselves as individuals. Team members reason thus: If I give 100 percent effort to the realization of Toyota's goals, then "we" will have a greater share of the market, sell more Camrys, acquire a reputation for quality, earn a bigger profit; hence, my job will be secure because there will be no need to cut back production or lay off workers. In other words, the employee through his or her own efforts has ensured a secure employment future and does not need to rely on the beneficence of Toyota to look after him or her. This rationale is congruent with and supported by what are considered to be the American values regarding the importance of individual effort in controlling personal destiny, and that hard work and effort are the means of success. It is fascinating that values of individualism have been woven into the community-of-fate ideology and, in fact, provide a primary support for it. Whether this represents a resolution of cognitive dissonance by American employees or a purposive Toyota adjustment of Japanese paternalistic ideology for individualistic American employees is unknown. What is clear is that job security, whatever its ideological foundation, is an important component in the individual-Toyota relationship.

The Lean Work Force Rationale The lean work force rationale is both a logical support for the belief that employees will not be laid off in depressed times and a rationale to justify heavy work loads. Every interviewee was conversant with the lean work force rationale. Martinez said,

> I did my research on Toyota. They haven't laid off since 1959 and that's because they only hired just enough people, trained people, that they could take care of in the lean years and the good years. Toyota in the United States is run the same way.

The relationship of the lean work force rationale to personal importance is emphasized by Fetalucci:

> I feel like I am important to them. I am important to them. One of the reasons for that is that they work with such a lean work force that if you miss, they have problems. They have to bring the team leader in to do your job and the group leader has to watch his job. You are important and I feel like I am important. It's a good job.

Griffin explained how the lean work force rationale justifies heavy work loads:

> Most of us, especially the engineers are really busy. We carry heavy workloads and we knew we were going to carry heavy workloads. Toyota policy is, we hire as few people as we can to get by, work a little harder in good times and then we don't have to worry about carrying around a lot of dead weight in slow periods.

The question that came to my mind during this analysis was why there was no skepticism toward this policy among my interviewees. No one questioned how many employees constitute a lean work force. No one challenged that Toyota's lean work force was too lean for personal comfort or organizational efficiency. No one questioned the inadvisability of Toyota supporting "dead weight" during depressed periods. All seemed to accept as a matter of fact that job security was ensured by the lean work force and that the job security so obtained was worth their present heavy work load and commitment. Undoubtedly, there are some TMM employees who do question these issues. It would be especially remarkable if there were not a few skeptics. Even so, I think it is significant that none of my interviewees were suspicious. Indeed, not one of them attempted to justify their belief in the lean work force rationale as they might if they were familiar with a contrary view presented by a disgruntled faction of employees. The lean work force was presented as a fact, not a belief.

Unquestioning belief in the lean work force rationale is partially explained by Toyota's past performance in Japan in regard to job security, but also by realizing that the lean work force rationale is rather appealing. It gives workers what appears to be a logical reason to be confident of their job security without resorting to simply trusting in Toyota's good intentions. Like the employee explanation for job security, the lean work force rationale places control for future destiny with the individual employee. It supports the notion that future security is earned, not a gift from a benevolent company, and that a heavy work load now is a type of deferred gratification for future reward.

The fact remains, however, that Toyota is not bound in any legal way by the lean work force rationale, or any employee rationale, to provide lifetime employment. Toyota has the power to lay off workers in depressed times and to lay off workers who are past their prime or less productive in some way. Toyota's past performance in Japan may not be indicative of how they will treat American employees. Squarely facing these facts makes it obvious that the only real basis for individual feelings of job security is trust that Toyota will abide by all the clauses of the tacit agreement. No wonder employees are so unquestioning of what appears to be "concrete, logical evidence" that Toyota will comply with the implicit contract they believe they have. For to challenge this "evidence" would reveal that the real basis of this agreement is trust, and that fact may be rather uncomfortable for American workers who are exposed daily to the news, even on their own Toyota television monitors, of violations of trust by other organizations and who have grown up at least partially believing that individual effort alone is sufficient to determine one's destiny.

This analysis has used the tacit agreement as a conceptual tool to help explain feelings of equity, open communications, bureaucratic flexibility, and job security as they exist at TMM. I have argued that these are important dimensions of the company team and have demonstrated how they are actualized, both in terms of organizational policies and structures, and in the perceptions of employees of TMM. These dimensions as a whole create and nurture the community-of-fate ideology and thus provide the mechanisms whereby the company-team concept can coordinate and motivate individual employees toward organizational goal achievement. Throughout the analysis it was apparent that Toyota faces challenges in actualizing and maintaining the company team. Among the most important challenges are mediating between the sometimes disparate and opposing definitions of equity held by different occupational groups of employees; preserving the benefits (in terms of implementing the company team) of bureaucratic flexibility without allowing it to degenerate into favoritism and inconsistency; and maintaining the amorphous trust that encourages employees to give Toyota the benefit of the doubt and encourages belief in major components of the Toyota ideology such as the lean work force rationale.

Importance of the Individual

One would expect that, in a Japanese organization oriented around the team concept, the last thing that would be encouraged is individual importance. Therefore, I was quite surprised when almost all Toyota team members interviewed told me how personally important Toyota made them feel. I will quote some of their comments to give you the flavor of the sentiments. Woulton, a team member, contrasted Toyota with other organizations at which she has worked. She said,

> I know how cold it can be sometimes. I'm not saying that Toyota could not be cold if it has to be. But I'm saying that I think they have more of a concern for their employees than other organizations. If an employee has a problem or something, they'll bend over backwards to do something about it.

An electrician, Raphael Martinez commented, "That's how Toyota stresses people's importance by saying, 'If you're not here, you'll be missed. You're needed here,' which is good. It's good knowing you're needed." Martinez describes how, when a salesperson comes to the plant to sell a product he (Martinez) uses, the salesperson must consult with him: "I never thought I'd be meeting with salesmen. He'll give me a price on a belt or something. I'll make the determination and recommend it to my group leader."

Another team member, Dave Fetalucci, reported it this way: "They trust their employees. I feel real confident that they do trust me. Basically I feel real good about my job. I feel like I'm important to them."

A group leader, Eric Klaske, described how Toyota treats team members:

> Toyota is fantastic. If it has a group of team members who don't like the way the parking lot is arranged, they'll change the parking lot. If they don't like the way the lunch room is working, they change the lunch room. I mean if there's a group consensus by the team members that something is perceived as not being as it should be, then Toyota tries to revise that, and you don't get that other places.

These feelings are reactions to an intricate mesh of Toyota programs. Obviously, many of the policies and programs already mentioned play a part in creating the perception of personal importance, such as bureaucratic flexibility, the lean work force rationale, a strong human resources department, open communications, and the leveling of status differentials between management and labor. Add to that list the employee training programs, the in-house child-care facility, and the employee assistance program (a benefit package that provides financial assistance for such services as psychiatric care and family counseling). While these programs appear logically to contribute to feelings of personal importance among employees, they did not loom large in interviewee comments. Training programs, other than the prepromotion classes, were mentioned only twice, both times by employees in the skilled trades who have more opportunity to utilize the programs. In these two cases, the training programs were a very significant component of the team members' positive attitude toward Toyota. Child care and the EAP programs were each referred to once. Only one interviewee, Bob Dottard, had need of child care, but he and his wife chose not to use it. Since these programs have so little direct impact on the lives of the particular sample I interviewed, they are not likely to be the organizational features that come readily to mind when they discuss Toyota. Still, I contend that they contribute to the overall ambiance that reinforces the message of personal importance, especially to team members.

Convincing employees of their personal importance is a critical link between the community-of-fate ideology and organizational goal achievement. It is not so remarkable that employees believe their personal welfare is tied to the success of the organization—even employees of the most callous and exploitive employers recognize that if their employer fails, they will lose personally and, while they may not gain equitably if their employer succeeds, at least they will keep their jobs. What is extraordinary, and much more difficult to achieve, is persuading employees that their individual performance directly affects organizational success. Especially in organizations with large numbers of employees and lifetime employment, it is understandable that employees might have trouble recognizing that their individual efforts actually have an impact on overall organizational performance. Policies and practices that demonstrate to employees that they are personally important to the organization might encourage them to assume personal responsibility for the organization's success. At TMM the message that individual efforts matter to organizational success is reinforced by provisions of the work team—for example, close supervision, small group norms and sanctions supporting organizational goal achievement, and kaizens. Taken together, TMM has created a powerful and compelling mechanism to elicit individual employee commitment to the achievement of organizational goals. As a summation of this section, I refer to a comment from the articulate Prichard:

> If the quality is good, if the product sells, they will be taken care of. In other words, they're not just being fools for working hard. People where I work now think they would be a fool to work hard because they wouldn't get anything for it. People at Toyota, most of them feel that if you work hard, do a good job, it's going to come back to you. Being able to foster that belief in people, to whatever degree it's true, is one of their biggest successes to this point and they're scared to death of losing it. As well they should be.

The Corporate Team

The primary purpose of this research was to construct a theory to assist in understanding work in a particular Japanese organization. During the course of the analysis, the work team and the company team emerged as the most obvious conceptual categories. It became clear, however, that limiting the analysis to these two units provided an incomplete picture. For even though it is true that employees interact directly only with their work team and the local company, the preconceptions that they bring to their interactions with TMM and the parameters within which TMM can act toward employees are determined by TMM's membership in the corporate team. The examination of corporate team will be limited to areas where it interfaces with the company team and the work team. In that regard, this will be a very incomplete presentation of corporate team, but the limitation is necessary to maintain the focus of this book.

The corporate team is a whole family of organizational entities who work cooperatively together and have a sense of belonging together. It includes TMM and the Toyota manufacturing facilities in Japan and elsewhere, financial and philantrophic organizations, research and development organizations, marketing divisions, and suppliers and subsidiaries of the corporate Toyota entities. The corporate team is referred to frequently in Toyota publications (i.e., the annual report, *Toyota Today*, and *Toyota Topics*). In *Toyota Topics*, a regular feature selects different suppliers and provides information regarding their contribution to the TMM effort. Within the corporate team Toyota manufacturers develop and nurture long-term relationships with their suppliers. The relationship between Toyota and its suppliers reaffirms to employees that Toyota is committed to long-term relationships. This is evident in Bob Dottard's description of an incident and what it meant to him:

> I remember once we were having trouble with the masking, with the type of masking we were getting. A team member went and asked them why they didn't dump this supplier and get another supplier. The answer she got back was, "Well, Toyota is kind of like a family. If you have problems with someone, you don't just throw them away. You try to work it out, find out why there's something wrong, take countermeasures and take the extra trouble." If you have

trouble on the line, or with a team member, you try to work
it out, come up with countermeasures.

Dottard explained that this event communicated to him that Toyota
would stick to its relationships, even when the association was problematic, and that Toyota has had successful experience in dealing with these
kinds of problems. The lesson, according to Dottard, is that Toyota does
not consider placing blame and terminating relationships as solutions
to problems, either with suppliers or, by extension, with employees. The
way business is conducted at the corporate level, at least in those areas
of which employees are aware, influences the internal climate of TMM.
    Another way in which the corporate team impacts the operations
of TMM and the company team is through the financial resources provided to TMM as a member of the Toyota corporate team. Mentioned
previously is the effect these resources have on the possibility of performance awards (bonuses) for employees. We must consider also the
probability that the existence of these resources, furnished by the family, as it were, decreases the pressure on TMM for the realization of
quick return on investment, or quick profit. Undoubtedly this allows
TMM to take a longer-term perspective in regard to profit making and
permits investment in such trust-enhancing and team-building organizational features as child-care facilities, wellness programs and facilities, personal-touch moneys, and bonuses that might be contraindicated
with the imperatives necessitated by a quick return on investments.
    In addition to financial resources, the corporate team provides
TMM with a technological legacy. That is to say, the way products are
manufactured at TMM is determined by the technology and processes
developed at other, older corporate manufacturing facilities. It is in this
respect that most American employees are consciously aware of being
exposed to the corporate team. Just-in-time processes, the kanban system of inventory control, the andon board and cord, and the kaizen system were all developed in Japan, as their names suggest. The unique
technology of the manufacturing processes at TMM has not been the
focus of this work. Enough has been written about Japanese manufacturing technology elsewhere. Important for our consideration is that
these features are part of the day-to-day reality of work at TMM and
that the prior success of this system undergirds the job security dimension of company team.

At least equally significant to TMM operations are the more intangible elements of corporate team such as its culture and history. The corporate history of financial success, which as indicated above, encourages employees to expect the future success of TMM, and the perception among employees that the corporation has a history of treating employees humanely, to use Barbara Woulton's terms, encourages employees to trust that they will be treated fairly. The Toyota philosophy and culture frequently referred to in this analysis originated in the corporate team. Further, the presence of Japanese trainers, and less so Japanese executives, lends an exotic element to TMM that is relished by many employees. The Japanese physical presence has diminished now with the return of most of the trainers to Japan. Nonetheless, American employees interviewed generally recalled positively their interactions with the Japanese. Some viewed it as a tremendous opportunity to have experienced another culture through their contact with Japanese trainers. An interesting perspective regarding the Japanese presence was described by Carol Ruddlehouse. I asked her about the difference between Toyota and places where she had worked previously:

> Just having people from another country to work with makes a big difference. I have worked with people from India, but most of the people I've worked with from India were at the university and then when I was with the government, I worked with an Indian person. It's still just different here, because the Indians had been in the United States for years and years, and the Japanese people are new here. Also when the Japanese people came here, they are the big time bosses. I know they are going to make or break my future.

Ruddlehouse was the first person I interviewed. After thinking about her comment I thought I had discovered an important aspect of Americans working in a Japanese organization. However, I never heard the same theme again. I had to ask several interviewees if TMM was really a Japanese company since up to that point they had provided no referents to the Japanese nature of TMM. After being asked, they proceeded to tell me about their work-related or social experiences with trainers, or Japanese executives, but no hint of perceptions of power or status differentials was apparent. Aside from Ruddlehouse's statements, the only

exception to this was from Tom Lemming, a team member, and Pat
Griffin, a specialist who told about what they thought was a major problem for upper-level American managers. Lemming said,

> From my personal point of view, Americans are led to
> believe that they (the Japanese) came over here and put up
> the money and they help train everybody and then put
> American management in place and went back home. But
> when you look at the breakdown of the managers, none of
> those managers has the authority to make a decision without
> consulting the Japanese coordinator. It's very frustrating
> and I see it in managers a lot.

The result of this, according to Lemming, is "I see them as blockers, to
screen or to block any decision being made without home office or
upper Japanese management knowing about it."

In a similar vein, Griffin reported:

> One thing I think is a real problem with TMM is that we
> have Japanese management and we have American manage-
> ment. Every manager has some Japanese counterpart and it
> seems that American managers go off and have their meet-
> ings and the Japanese go off and have their meetings. Toy-
> ota is organized so that a Japanese manager and an Ameri-
> can manager are side by side and this goes right up the
> ladder to the president. Well that means that a lot of times
> when there's any kind of disagreement between an Ameri-
> can and a Japanese manager, it has to go all the way up to
> the president to get resolved. I don't think there's enough
> interaction between the Japanese managers and the Ameri-
> can managers in some departments.

At the end of the interview Griffin told me that the dual Japanese and
American hierarchy was the single most important thing he wanted to
communicate to me. Griffin is referring to the dual Japanese and American hierarchy as a hindrance to organizational efficiency. Lemming,
however, views it as a check on the power of American managers that
can be frustrating personally to managers but—more importantly in his

opinion—reveals the permanent junior status of the Americans in the company-team enterprise. The two upper-level managers interviewed did not express these concerns. Still Griffin and Lemming, even though not personally in these situations themselves, present a plausible position. Support for this position is common in the literature. For example, Malcolm Trevor's (1983) study of Japanese multi-nationals in Great Britain reports that British nationals were allowed only a limited role in the Japanese organizations studied. Sethi and colleagues (1984) identifiy this as a major problem faced by American nationals employed by Japanese organizations in the United States. Moreover, support is provided by an examination of the structure of the organization in that Toyota has placed Japanese nationals in all executive positions except in human resources and public relations departments.

If indeed this is a credible concern, the question then is, Why are the American managers, who were formally or informally interviewed, unaware or unconcerned about this issue? It could be that they are well aware of these problems but out of loyalty to Toyota, or because in talking to me they assumed their official spokesperson-for-Toyota role, they did not want to reveal any potential Toyota faults. It could be that it is not a major problem in their particular departments or at this particular time in their departments. On the other hand, it is possible that there is a component of the Toyota ideology presented to American managers that anticipated this concern and attempts to belay any problems associated with it. The rationale that makes the dual hierarchy and a power differential between Japanese and American managers palatable to the Americans is apparent when we examine Huntington's portrayal of himself as a student of the Toyota system of management.

That upper-level American managers see themselves as students learning the Toyota philosophy and methods was expressed by both Huntington and Axelrod, echoed in a published interview with the American senior vice president, and was frequently communicated to me in informal conversations with American managers. This perspective encourages the attitude among managers that they are not yet ready to assume a more powerful role in the organization. Implicit to the acceptance of this rationale is the belief that eventually the American managers, if they are patient and learn their lessons, will be ready and, then, the reins will be handed over to them.

Undoubtedly, a factor in the acceptance of this rationale is the relative autonomy managers have, vis-à-vis Toyota, due to their advantageous position in the labor market, at least as compared to team members. Upper-level and most middle managers have had prior experience in their area of expertise. They have demonstrated, by moving to take a job with Toyota, that they are not locked into a particular geographic area and will move for the right opportunity. For some of these managers, learning Japanese management, Toyota style, will be rewarded, if not by promotion within the company then by enhanced employability with other companies. In this light, temporarily assuming the student role is a no-lose situation for a manager who is willing to leave Toyota and relocate. Roger Slider who is a lower-level manager aspiring to move to the upper levels, summed up this logic:

> My original thinking when I came with Toyota was that I would learn whatever I could about robotics and so on, wait five years, and then look around me and see if I like what's going on. If I'm not making headway, I'll look elsewhere. My personal philosophy is that if I don't get promoted in so many years, then I will resumé somewhere else. I told my managers, "I came here for one thing, to advance." I don't mind putting in extra overtime to do it, giving 100 percent.

The attitude of enhanced employability may be the personal trump card that allows American managers to counsel patience and assume a learning posture. An intriguing facet of the attitude of enhanced employability is that it is in contradiction to the community-of-fate ideology and therefore, must be kept private and perhaps even unspoken by the managers who are charged with the responsibility of generating an atmosphere conducive to the company team. Slider, in the above quote, does not give the impression that he will sink or swim in accordance with the fate of TMM. Rather he is expressing the notion that he works hard to better his own personal fate and that it is possible that his best interests and those of Toyota may be in conflict. Recognition of the dual hierarchy and the attitude of enhanced employability raises the specter that not only are Japanese and Americans not in this thing together, in the community-of-fate sense, but also that American managers and American team members, because of different universes of opportunity, do not share a common fate. Clearly, credible articulation of these possibilities could be extremely destructive to the company team. In retrospect, Ruddlehouse's perception of the Japanese as the "big bosses" may have been prescient.

Slider is the only manager, indeed, the only employee, who spoke bluntly about enhanced employability. Hence, it is possible that this attitude represents an extreme minority opinion among the managers. The factors that discourage expression or even recognition of this attitude are quite pervasive and therefore, determining the extent of it is difficult. Nonetheless, at some point in time American managers will grow impatient with the student role and see themselves as ready to assume responsibility for TMM's operations. When that occurs, if managers perceive an inequitable sharing of power and a ceiling on advancement opportunity, enhanced employability will provide them with the independence to sever their ties with Toyota. The damage sustained to the company team in the process may be irreparable.

The irony is that the perspective of enhanced employability among some American managers may be a crucial element limiting their access to the power heights. In the literature, Japanese managers are purported to believe that Americans have a mercenary attitude toward work and their employer (Sethi, Namiki, and Swanson 1984; Tsurumi 1978; Howard, Shudo, and Umeshima 1983; Omens, Jenner, and Beatty 1987). According to these authors a most important component among upper-level managers in Japanese organizations is company loyalty. If American managers threaten to walk with their newly acquired knowledge of the Toyota system, their company loyalty will certainly be suspect, thereby limiting the very opportunities for advancement they hoped to promote.

To summarize: the corporate team promotes and sustains the company team through its history, culture, philosophy, production technology, economic resources, and exotic ambiance. In this way corporate team impinges on the day-to-day dimensions of reality of TMM employees. On the other hand, the corporate team represents a substantial potential threat to the company team. If the Japaneseness of the corporate team comes to be perceived as a hindrance to Americans in their realization of full equitable membership in the company team, then the fragile atmosphere of trust necessary for acceptance of the community-of-fate ideology will be severely threatened.

5

ON RESTRICTION

The issue of work-related injury and illness looms large in the day-to-day reality of production workers and supervisors at TMM. Workers who are injured or ill and cannot perform their normal job, but do not require rehabilitation at the hospital or home, are placed "on restriction." All production employees interviewed were either on restriction, had been on restriction, had their jobs affected by teammates on restriction, were concerned with preventing the injuries that might lead to restriction, had developed or implemented kaizens to prevent workplace injury, participated in one way or another in the safety programs to prevent accidents or illness, or as supervisors, identified injuries and illness as one of the most problematic areas of their jobs. The only workers who appeared relatively unaffected by injury, and the concomitant restriction, were employees engaged in administrative functions such as the purchasing and accounting staffs.

On restriction deserves our attention and a chapter of its own because it was identified by interviewees as a significant element of their work life and because it represents an important challenge to the philosophy of Team Toyota both at the work-team and company-team level. Briefly, a significant number of restricted workers causes a breakdown in work teams and leads to uncertainty about the veracity of major components of the company-team ideology. Conversely, properly functioning work teams are a strong preventative of workplace injury and, according to the company-team ideology, Toyota, concerned with the welfare of all employees, will marshal the resources and leadership to effectively address the problem.

Definition

Before embarking on the meaning of the concept *on restriction* in its official and colloquial uses, it is necessary to clarify the distinction

119

between workplace disability and on restriction, and justify my choice of on restriction as the more important of the two phenomena. To begin, workplace disability is officially subdivided into two categories—injuries and illnesses. Workplace injuries occur because of discrete events or accidents. Examples include broken bones, lacerations, and sprains. Illnesses, on the other hand, are the culmination of normal job activities performed over and over again. Repetitive motion illness—called "carpal tunnel" by Toyota employees—back problems, hearing damage, and brown and black lung are types of workplace illnesses. Since the words *injury* and *illness* have precise and different meanings within the literature, I have decided to use the term *disability* in this study as the broader category that includes both work-related injuries and illnesses.

Regardless of the kind of disability involved, the disability itself refers only to physical symptoms, medical diagnosis, and procedures. The concept *on restriction* as it is developed here, however, includes all the salient social features and consequences of the disability that pertain to the work place. Further support for the distinction between disability and on restriction is provided by the fact that team members can be seriously disabled but, by personal choice, not on restriction; and that in the eyes of coworkers and supervisors, team members can be on restriction without being seriously disabled. Although the two are related to each other, the focus here on workplace disabilities is only within the context of on restriction.

On restriction refers to the situation of workers who are prohibited from performing their regular job due to a work-related disability, but who come to work and receive full pay. Placing disabled workers on restriction is intended to provide them with relief from their former jobs to facilitate healing and at the same time maintain their wages at preinjury levels. To be officially classified as restricted a worker must have a work-related injury or illness, provide medical verification of the disability and the need for a restricted schedule, and receive the official sanction of the human resources department. Medical verification can be provided by either a Toyota physician or a personal physician. Team members on restriction may be assigned a less demanding job in their group or in some other department, or they may perform the duties normally assigned to their team leader. The team leader replacement is often necessary, since she or he is usually the person filling in for the restricted worker. To the extent that the team leader is not as proficient at the job as the team member on restriction, other team members will have to pick up the slack. In a lean work force situation, there are no extra workers to fill in for missing or restricted team members.

Other provisions of restriction are that restricted employees cannot transfer, work overtime, or be promoted. They must be medically certified as able to perform their preinjury job before being taken off restriction and, therefore, before promotion, transfer, or overtime are possible. Employees may not stay on restriction forever. After a period of time, they must be able to perform their preinjury job or be classified as permanently disabled, in which case their pay is decreased by an amount determined by their ability to perform a different job and is covered by workers' compensation. Officially these decisions are made by the medical professionals involved, in consultation with the disabled employee and the human resources department. Sometimes disabilities require workers to rehabilitate at home. Under these circumstances pay is reduced by 35 percent for up to a period of two years, after which they would be classified as permanently disabled. After one week at home, they are paid through workers' compensation. If the injury was immediately diagnosed as permanently disabling, they would be moved to workers' compensation sooner. Individual employees may move on and off restriction, and back and forth between work and home while disabled, depending on their particular situation. Due to the pay differential, disabled employees prefer restriction to being confined to home or hospital with a disability, and since restriction has a limited but flexible time span, employees are encouraged to move from restriction back to their regular job or face being categorized as permanently disabled, losing their job, with a probable reduction in compensation.

Extent of Disabilities and On Restriction

The injury and illness rate for TMM was relatively high according to an article in the *Lexington Herald Leader* (April 1990). In the article Toyota reported that the episodes requiring medical attention beyond first aid and resulting in lost workdays for 1989 (the first full year in production) were higher than the industry average—14.92 per 100 production workers compared to 12.8 per 100 workers industry wide. Toyota is quick to point out that their rate is figured using production workers only as the base, while the industry rate is based on all employees of the organization. Production work is more hazardous in terms of injuries than is office or administrative work. Therefore, counting office and administrative workers in the base deflates the injury rate present in the industry. Also since national 1989 injury and illness figures were not yet available at the time of the article, the comparison being made is between the TMM 1989 rate and the national 1987 rate. Thus, whether the Toyota rates are actually higher than national levels is unknown for comparable time periods. Still workers and Toyota officials alike identify work-related disabilities as a serious problem.

In 1990, the TMM rate for recordable cases decreased by 40 percent, but the rate of lost work increased by 4 percent for a total combined rate in 1990 of 12.12 per 100 workers (*Toyota Topics,* February 1991). This represents a decline from 1989 and, according to the Toyota publication, a significantly lower rate than the industry average of 32.47 for 1990. No explanation is provided for the apparent astronomical increase in industry rate from 1989 to 1990. According to the newspaper article mentioned previously, Toyota believes the most troubling kind of work-related disability is repetitive motion illness. This is caused by performing the same stressful arm or hand motions over and over. The incidence of repetitive motion illness is less than some other types of disabilities at 18 percent of all reportable cases. Lacerations accounted for 30 percent, strains 20 percent, and miscellaneous injuries such as foreign objects in the eye, fractures, and abrasions make up the rest. The actual incidence of repetitive motion illness is higher than reported figures, but I will return to this issue later. Despite its apparent lower frequency, the reason for the concern about repetitive motion illness is that it is more difficult to predict, prevent, and cure than other types of disabilities that occur at TMM.

From the point of view of production workers, repetitive motion illness is also more problematic than other kinds of injury. This is so because it may take a long time to appear and it is a less obvious injury than cutting a hand or breaking a leg. Hence, even though workers with repetitive motion injury may be in great pain and permanently disabled as a result of their illness, they may not themselves be able to distinguish between the onset of repetitive motion illness and the normal fatigue, aches, and pains associated with demanding physical labor. These same characteristics, gradual onset and lack of visual signs, make victims of repetitive motion injury more likely to be suspected of malingering. We would expect that among the thousands of TMM employees, a few probably are malingering. This adds just enough credibility to the charge of malingering that some employees with repetitive motion illness who are on restriction perceive more ambiguity and less support from coworkers and supervisors than do employees with other types of disabilities. This is a particularly cogent factor in a team work situation where other team members may have to carry some portion of the restricted team member's work load in addition to their own. While medical confirmation of the disability caused by repetitive motion illness and Toyota's treatment of repetitive motion illness as a serious injury does help to legitimate the condition as a "real disability" in the eyes of employees, enough ambiguity remains to make repetitive motion illness a socially uncomfortable reason to be on restriction.

Given the debilitating quality of repetitive motion illness and the slowness with which it heals, the percentage of people on restriction who have repetitive motion illness at any given time is likely to be higher than we might expect from the comparison of the rates of different types of worker injury and illness provided above. Every interviewee except one who talked about on restriction was referring to workers with repetitive motion illness. In the exception it was hard to determine what factors caused the restriction of the worker discussed by the interviewee. For the interviewees, being on restriction was synonymous with having repetitive motion illness.

We must be careful not to equate the disability rate with the restriction rate, which is unknown outside official Toyota channels. The disability rate counts episodes and medical visits and not disabled employees. Employees on restriction would be counted in the disability rate since a medical visit is required, but one restricted employee may be counted over and over again during the year as he or she revisits the physician, moving on and off restriction. Similarly, some employees who require medical attention or even miss work for health reasons may not be placed on restriction. A lacerated thigh or ear that required stitches and medical attention—and, thus, counted as an official disability incident—may pose no hindrance to normal job performance or necessitate recuperative time on restriction. Clearly, these are distinct, albeit related, phenomena.

In the absence of hard statistics regarding the restriction rate, we must turn to employee perceptions of the rate to gain a sense of the magnitude of the situation. Using this technique, it must be noted that employee perceptions will reflect most prominently the situation in their particular group, or at most department, and that my sample is small and nonrandom. Therefore reports may or may not be a true picture of restriction rates plant wide. Even with these cautions in mind, some facts are fairly certain. Certain departments have a much higher rate than others, and regardless of the rate, workers view restriction as a serious problem. I interviewed five employees of the department generally regarded as the most hazardous. One of those employees did not mention specific instances of injury or restriction but commented frequently on the need for safety programs, kaizens, and redesigning equipment. All the others saw restriction as a prominent feature of their work. A team member noted: "We sat down one day and counted all the people who worked first shift. Of those, 65 percent had been injured and were on and off restriction continuously." Another team member said,

> The last I heard, there were six restricted in our area. (I am not sure what the meaning of area is here. Usually employees used area to describe the place where their group worked. I am assuming therefore that the area contains between sixteen and forty employees, or one or two groups.) I don't know what their restrictions consist of, each different ones. But there are six restricted people on the night shift right now too. It finally caught up with them. They started out and they were fine and we were all restricted at that point in time. Now it caught up with them.

Although this department is probably the most extreme case at TMM, it demonstrates that restriction is a severe problem for this department, while probably a less important problem elsewhere.

Factors Related to On Restriction

On restriction is one method TMM has developed to respond to the problem of work-related injury and illness. Within this context on restriction is a structural feature of the organization with an official definition and terms. However, an understanding of this concept acquired from the official definitions is meager in comparison with the meanings it has to workers. On restriction impacts almost every aspect of work life for Toyota production employees from the designation of certain departments as high-risk areas to be avoided, to self-imposed restriction in choice of future jobs necessitated by the effects of unreported injuries. In order to explore this complex of meaning I will elaborate those factors related to on restriction. These include the job performed, the type and severity of the disability, and worker attitudes.

The Vicious Cycle of Workplace Injury

Employees in certain job categories are more likely to become disabled (irrespective of whether they are on restriction) than employees in other categories. Partially this is a result of the more dangerous nature of certain jobs, but also it occurs because departments can get caught in a loop where the presence of disabilities increases the likelihood of future disabilities. As noted above, certain departments within production seem to contain a larger percentage of the disability prone jobs than other departments. These departments have developed a reputation for unpleasant and potentially hazardous work, although even in those areas some job are "cushier" than others. Initially when a team member joins the organization only the "luck of the draw" determines whether s/he will be placed in one of the disability prone jobs. After initial placement, team members avoid transfers to the jobs and departments reputed to "beat up people" and many team members in those areas attempt to be transferred out as soon as possible.

   This creates a vicious cycle: since experienced workers avoid those areas, the worst jobs are more likely to be performed by inexperienced teams. Inexperienced team members will be unable to rotate fully until all team members become proficient on each others' jobs. A continuous supply of inexperienced team members makes cross training a continuous process that, under the demands of a lean work force and a rapid work pace, often becomes a prohibitive luxury. Additionally, inexperienced team members are not as likely to be in good physical condition as are experienced team members who have become accustomed to the rather strenuous demands of the jobs. These contingencies—lack of rotation and lack of physical conditioning—increase the probability of disability for healthy team members, especially in jobs that require repetitive movements. Thus, a high number of restricted people in a group will create a situation where injuries, illness, and on restriction will be more likely to occur in the remaining healthy team members.

Johnson's department has a reputation for jobs that "beat people up." He says the situation sometimes is so bad that the team leaders cannot fill in for all the restricted workers. Team members have to be borrowed from other sections, temporary employees must be brought in, or summer replacement employees are utilized to keep the line running. The constant threat of disability, turnover of team members, recruitment of team leaders to fill in line jobs, and the subsequent chaotic and crisis temperament this creates demoralizes team members and destabilizes work teams operating in the hazardous departments.

The Significance of Carpal Tunnel Syndrome

Another important consideration regarding who goes on restriction is the type and severity of the disability. Some disabilities obviously require placing the affected team member on restriction. In the case of injuries such as broken bones or severe lacerations, there is no ambiguity. The team member has little discretion in the matter. On the other hand, afflicted workers, lay observers, and even medical professionals have more difficulty determining the severity of some other injuries, and in particular, illnesses. For example, at what point in the progression of repetitive motion illness does it become severe enough to require restriction? One would think that when the pain becomes excruciating, or when the affected wrist, fingers, or elbow lose all feeling, or when there is a possibility of permanent disability, the worker should be restricted. Consider, however, that each individual's pain threshold is different, that other contextual elements might encourage the worker to exaggerate or ignore their symptoms, that on a different job the problem may not bother them at all, and that the medical diagnosis and treatment of carpal tunnel are inexact and open to various interpretations. In this light one begins to see that for some disabilities, and especially illnesses, much more is involved in the determination of going on restriction than merely the severity of the physical condition. Workers' attitudes, the social context, and individual assessment of the situation become equally significant.

Greenly, who has been diagnosed as having carpal tunnel syndrome, was hired in the first wave of new employees and with the luck of the draw, she says, landed in a group where the jobs are prone to creating repetitive motion illness. She has been on and off restriction several times. Through her account we can see the conflict faced by team members with carpal tunnel in regard to restriction:

> I told my doctor unless you take me off restriction, there's no way I can ever get out. So he did take me off restriction other than to rotate every two hours. I waited all year and nothing. I waited as long as I could stand it and no transfer. As soon as I went back on restriction, about two weeks later, Jacob transferred out of our area. It was a job I could have done and I would have had a good chance for it.

Since workers on restriction cannot transfer out of their group, some of them feel that they are trapped in their group until they are well enough to perform their job. Johnson supported this theme when he told about a teammate:

> He had been having awful hand problems, but he would not go up front. (*Up front* is a term used to refer to going to the administration. In this case it means getting official recognition for the injury and being placed on restriction.) His group leader talked him into going up front. He did, and as soon as he did, they restricted him. They told him he could not transfer. He was due for a transfer and they promised him a certain job he thought would not hurt his hand. Soon after he was restricted, they came over and offered the job to Joyce. You really get between a rock and a hard place when you have carpal tunnel.

Clearly, being unable to transfer when restricted offers a potent disincentive to going up front with an injury like carpal tunnel. Conversely, as Greenly pointed out, there is more involved here than keeping a job:

> After I did that same job all day long, as I was driving home, my hand started to draw up. I thought, now see, cripple time. Here we go. So I went home and put ice on it and it eased back up. I kept ice on it until 9:30 that night and it did ease

back up. But I do think I'm getting close and I don't want to permanently damage my arms to where I can't work some-where else or do something else.

The Significance of Work Team

Team members react to workers who are on restriction. Greenly wears a wrist brace to protect her wrist from further damage. Toyota encour-ages team members to wear a brace if they or their doctor feel it will help. Greenly had hers on during the interview. But she said she takes it off when she goes to the Toyota cafeteria. The reason, she said, is because some workers make negative remarks about the "cry babies" who wear wrist braces. She maintains that there is a stigma associated with having carpal tunnel and the brace is the outward symbol of the stigma. Only one interviewee, a group leader, supported the notion that being on restriction for carpal tunnel carried negative connotations. Klaske contended that one of the hardest parts of his job is determining who is malingering and who is not. Actually, he pointed out, the medi-cal staff makes this determination, but "it knaws on him" because he believes that some employees are overreacting, looking for the easy job, or need to learn to tolerate a little discomfort. He thought Toyota is too unquestioning and accepting and, thus, is being taken advantage of by a few unscrupulous employees. Even though there is no official weight given to his opinions about the legitimacy of team members' medical problems, his attitudes undoubtedly affect his restricted team members and their coworkers in some way. Plus he can encourage or discourage them from going up front with a suspected illness or injury.

More often I heard comments from employees like that expressed by Perkins who said, "If you know you're hurting, they want you to get the problem resolved. They offer a little wrist band that will try to keep your tunnel area straight. A lot of people just keep going on and on and on until it's too late though."

According to Woulton,

There are some people who are concerned about carpal tun-nel. But there are some people who aren't injured and they're not concerned. These injuries bother me. I've seen what it does to people. It's not just something that they have to live with the rest of their lives, it affects their livelihood. It's not that the others don't care, but they depend on management to do something about it. There are people on the line who were previously injured and some are working with injuries. They're still going to be there until they can't make it any more. A lot of them don't want to go to medical and be put on restriction and sent home. They'll stay in there and work until they die.

A similar sympathetic theme came from Prichard, who, like Klaske, identifies injuries and on restriction as one of the most problematic aspects of supervising employees. However, Prichard, while recognizing the possibility of malingering, was more concerned that his actions or inaction may have caused the team member's injury. If he had just not assigned that person to that particular job; if he had just worked out a different rotation schedule; if he had just kaizened a change in the job to make it less damaging; if he had just recognized the symptoms earlier and encouraged the team member to go up front the injury might have been avoided, or caught before it became permanently disabling.

A manager Dave Axelrod described positively the work attitudes of restricted employees who come to work and pitch in and do whatever they can. He told about restricted employees painting lines, sweeping up, stocking parts for other workers, and so on. Axelrod believes the willingness of restricted workers to do these menial tasks demonstrates the inordinate levels of work ethic present among team members. He does not believe the majority of restricted workers are malingering.

What are we to make of these conflicting accounts of the stigma attached to being on restriction? We should not be surprised to find that both perspectives are correct, at least to a degree. Perhaps Greenly is overly sensitive or even slightly paranoid about being on restriction. Still, it is certainly possible that at least a minority of team members stigmatize and maybe even scapegoat restricted employees, for restricted team members do pose problems for their teammates and supervisors. Restricted teammates are a constant reminder of the danger of the jobs performed by production workers. A significant rate of restriction in a group increases the work load of the healthy members, demoralizes the group, and probably actually increases the hazards to other members of the group.

With this as a backdrop, let us consider how the dynamics of small groups might affect this issue. If Toyota is at all successful in creating work teams, team members will feel loyalty and even some affection for teammates. An injured team member knows that his or her team-mates must bear an increased work load if he or she is restricted. There-fore, going on restriction would involve feelings of guilt in the disabled team member. In turn, team members would be positively motivated to work with pain in order not to let teammates down and to maintain a good reputation with teammates. On the negative side, work teams might develop techniques to sanction members who are perceived as purposely avoiding the cooperative reciprocity of team work. In small groups these usually include negative comments and social avoidance, but may escalate to ostracism and even physical assault. There was no indication in my interviews of any of the more severe small-group sanc-tions. However, sanctioning actions may explain some of the negative stigma described by Greenly. In other words, if a stigma exists it may be a sanction used by team members to enforce the norm that all team members should carry their weight in the team effort. Under these con-ditions the small-group norms will encourage a disabled worker to per-form his or her normal job as long as possible before going up front with the problem and will encourage him or her to spend as little time as pos-sible on restriction.

Self-Imposed Restriction: A Personal Calculation

All of these factors—the inability to transfer while on restriction, the likelihood of a better job through transfer, one's personal capacity to endure pain, the potential of permanent damage, possible stigma, and loyalty to one's team—enter into the calculation made by team mem-bers with carpal tunnel or a similar disability before a decision is made regarding going up front. If a disabled worker decides not to go up front, she or he will be subject to self-imposed restriction and will use various strategies to be able to perform the job while disabled, hoping to hang on until a transfer to a better job occurs. Workers who utilize self-imposed restriction are restricted in the sense that they can only con-sider transfers to jobs that will cause no further damage to their bodies. When it is learned that a particular job may be open, a worker on self-imposed restriction will gain information about that job, assess its hazardous qualities, and decide whether to go after the job or take the transfer if it is offered. Greenly relayed an account of a teammate who chose self-imposed restriction:

> Even though he is not restricted by Toyota because he won't go up front, he's still restricted because he knows that his hands won't be able to take it. So when they offered him the transfer and they said he probably would be holding a gun all day long, he decided to stay where he was. At least where we are he doesn't hold a gun the total day.

Self-imposed restriction provides the benefit to disabled workers of the possibility of being transferred, albeit to a more circumspect number of jobs than if they were not injured. Additionally team members on self-imposed restriction do not face the possible stigma and guilt of being on restriction and, indeed, may even enhance their prestige with teammates as they continue to carry their work load even while hurting. The price of these benefits are working with pain and the potential threat of permanent disability.

Disabled workers on self-imposed restriction utilize various strategies to lessen pain and allay fears about permanent disability while performing their normal job. Strategies involve learning how to "shoot the gun with the other hand," how to use other fingers to operate equipment, how to substitute damaging wrist or elbow movements with less stressful positions or movements, wearing protective braces or clothing, designing lighter tools, or redesigning jobs and equipment. Greenly said, "If your right hand hurts, use your left hand and then trade back and forth. Shoot with a different finger. Shoot with all your hand or use a glove." However, sometimes the strategy can backfire. Greenly told how she advised another worker to learn to shoot with his left hand. Later when she asked him if he was trying to do it, he told her "No, I want one good hand." She continued, "I started laughing. I thought yeah, I guess that's probably true. I should have thought of that."

Furthermore, workers have devised home remedies and have altered off-work activities to assist them in performing their normal jobs. Recall that Greenly soaked her wrist in ice for hours after going home so that the next day she could execute her duties. She also reported that in order to minimize the strain to her wrists, she wears the wrist brace all the time and engages in little or no off-work activities that require wrist movement. She no longer does housework, goes bowling, or types letters. That way, she reasons, she will be resting her wrists as much as possible so that they can better withstand the pressure of her job. The off-work limitation of activities and the application of home remedies to relieve strain are important worker-generated strategies and reveal an extraordinary desire to stay off restriction and to remain employed by Toyota. The frequency of this last category of strategies is unknown and may be fairly limited. Even so, that it is present among even a few disabled team members is remarkable.

Many of the at-work strategies are suggested by Toyota and/or require Toyota input. Toyota involvement takes the form of safety representatives who advise workers regarding the prevention of disability or further disability, and whose function it is to assist in identifying the cause and solution of safety problems; maintenance teams assigned to implement safety kaizens; regular group safety meetings; and safety slogans, signs, contests, and special programs. Toyota safety efforts are clearly broader than just assisting disabled workers to perform their normal job. In fact, TMM would deny that it encourages disabled workers to remain at their jobs. Still as noted previously there is always ambiguity between the onset of some illnesses and the "normal" aches and pains of performing demanding physical labor. By encouraging workers with normal aches and pains to find ways to cope while performing the job, Toyota is, in the same process, assisting disabled workers in effectively executing self-imposed restriction, at least temporarily.

Implications for the Company Team

One consequence of underreporting and working with pain is that the hazardous characteristics of the particular job that caused the disability may not be addressed as a serious problem and, thus, will not be corrected. No matter how aggressively Toyota attempts to prevent disabilities, unless it considers the disincentives involved in employees' decisions to report the problem, its efforts will be less than totally successful. A certain amount of disincentive to reporting may be viewed by Toyota as advantageous in that it discourages false reporting and keeps experienced workers on the line as long as possible. Even so, Toyota must be perceived by employees as vigorous in its efforts to prevent workplace disability, as sincere in encouraging workers to report injuries and illnesses, and as trusting of employees when they do report problems. Less than that would undermine the ideology of the company team that Toyota has so carefully cultivated.

The central tenet in the company-team ideology is the notion that "we are all in this together," that all classes of employees, and indeed the owners and employees, "share a common fate." If production team members came to believe that it is their particular and exclusive fate to face a significant risk of maiming and permanent disability, and that Toyota, managers, and owners, are unconcerned about, contributing to, or benefiting from this situation, then the community of fate is fractured along the traditional adversarial lines of workers and managers, owners. According to Greenly and Johnson, some employees in the departments reputed to be the most hazardous express just these sentiments. Greenly herself states that Toyota is not doing enough to eliminate hazardous situations, especially those that cause repetitive motion illness. It is as if, she said, they do not take her complaints seriously. Work-related injury is only one reason for the low morale in these areas. Still, the extent of the threat to the company team is indicated by the perception among interviewees that employees in these departments are more sympathetic to unionization than are employees in the rest of the plant.

Toyota policies make more sense if we realize that one of their primary goals is to sustain the company-team ideology. In Klaske's opinion Toyota is too soft and believing of employees with complaints about the hazards of their jobs. He said as a result Toyota is sometimes taken advantage of. If this is true, Toyota's response may be to consider a few false reports and some malingering as the price it is willing to pay for maintaining the trust necessary to the company team. At least in part this strategy is successful in creating the impression among the majority of my informants that Toyota is sincere in its efforts to prevent injury, not just for economic reasons but also because of the pain and hardships it causes individual workers. Some team members described their understanding of Toyota policy in this matter: "They want you to get help if you hurt," "They want you to wear the wrist brace if there is any possibility of carpal tunnel." A maintenance worker reported, "The last kaizen I implemented was a safety measure. Those get high priority. If somebody was in danger or somebody was hurt or straining something, those kaizens get the highest priority."

The story of the permanently disabled worker who was retrained and placed in a less physically demanding job served as a sign to informants that Toyota cares about disabled workers.

A viable company-team ideology, by its very nature, is a powerful disincentive to reporting injury. The potential of stigma associated with restriction, and the feelings of loyalty to teammates that discourage reporting injuries, are more likely to occur in an atmosphere where most workers believe their interests are congruent with Toyota interests. This atmosphere can only be sustained if the workers believe that Toyota is concerned about their safety. Thus, the costs of carrying a few malingering workers on restriction in order to maintain the company-team ideology may be more appropriately viewed as an investment in malingering prevention. Toyota appears to be the defender of injured workers and the negative factors that discourage reporting appear to be beyond the control of Toyota.

What is seldom considered by interviewees in this complicated issue of workplace disability are two factors that Toyota does control. The first factor is referred to by Toyota spokespersons in public comments regarding the injury and illness rates of Toyota. By intention, Toyota hired highly motivated, highly educated, inexperienced workers. The benefits of this policy to Toyota have been elaborated elsewhere. One of the disadvantages is that inexperienced workers who are not accustomed to physically demanding work are more likely to be disabled. If this is the major reason for the high injury and illness rate, we would expect that the rates will go down as the work force gains experience, becomes more physically fit, and attrition removes the weakest workers. The decrease in the 1990 rate is an indication that the low experience level of the work force may have been a contributing factor in the initial high rates. However, a second factor must also be considered. The lean work force policy may enhance the disabilities, under reporting, and the vicious cycle quality of disabilities in some departments. It seemed as if the workers interviewed were so committed to the lean work force rationale that they did not see its possible negative impact on workplace health and safety. Nobody said the reason for the injuries is that "we do not have enough workers to do the jobs safely." Nobody said "the solution is to hire more workers." Suggested solutions were proper job rotation, redesigning jobs, and changing equipment, in that order. Perhaps the reason the lean work force was not mentioned is because it in truth is not related to the disabilities, or because my sample was inordinately satisfied with Toyota and generally naive, or because the allure of security offered by the lean work force rationale prohibits critical examination of its implications. If the disability rate does not continue to decline as the work force becomes more experienced, and with aggressive management efforts toward prevention, then the lean work force rationale may be subjected to more scrutiny as it becomes a more apparent causal factor.

Medical Confirmation

One final element needs to be evaluated regarding on restriction and that is the medical certification of the disability. One might think we have finally arrived at the scientific, objective component of the restriction equation that will separate the truly injured from those faking, that will prevent the disabled from returning to the job before they are healed, and that will protect the disabled from working themselves into permanent disability. Not so. It is not my purpose to dwell on the inexact nature of medical diagnosis and treatment. My point is that social and personal contingencies influence medical decisions except in simple, clearcut cases with distinct physical evidence such as broken legs. Different physicians may disagree on the diagnosis and treatment of the same condition in the same patient depending on their personal orientation, their opinion of the patient, and the circumstances faced by the patient. Since this is rather common knowledge, a person may shop around for a physician until she or he finds one with an acceptable opinion. Company doctors (referring to doctors employed by organizations to treat the organization's employees) are frequently viewed as having a conflict of interest. People sometimes ask, Whose interests are company physicians most concerned with, the employees who might be sick or injured or the employer who needs bodies on the line? These physicians, rightly or wrongly, are often suspected of not being sympathetic to the workers' needs. Both of these circumstances hint at the implications of social meanings on the seemingly objective medical confirmation necessary to be restricted. In the first case, support is provided for those who think a particular person is not seriously ill even though that person has obtained medical verification from a personal physician. The second situation discredits the assurances provided by Toyota physicians that an injury is not serious. At least one interviewee expressed the above sentiment about the Toyota physicians. He said that whenever employees go to see them for problems related to repetitive motion illness, they have to endure "the speel," a sermon about learning to withstand a little pain or hints about avoiding the problem in the future. The implication is that the complaint is not being taken seriously and/or that the fault lies with the injured worker. Greenly sees a personal physician because she feels he will protect her health more diligently than the Toyota physicians might. Even so, he consented to her request to be taken off restriction so that she would have the possibility of transferring. Toyota usually accepts the decisions of personal physicians in regard to restriction. By allowing an "objective" third party into the restriction determination, Toyota is seen as trusting in the employee's choice of physician and as genuinely interested in safeguarding the employee's health. This position is consistent with the argument presented previously that one of the Toyota's main goals in its policies toward workplace disability is the maintenance, even the enhancement, of the company team, and not the ferreting out of malingering employees. In any case, this task is more effectively accomplished by individual teammates who are committed members of the Toyota team than by miserly appearing policies.

We can see that medical confirmation, while a necessary component in the restriction equation, is not sufficient. Depending on the social and personal circumstances of the disabled worker, medical diagnosis may be sought or not sought, viewed as legitimating the disability in the perceptions of coworkers or not legitimating it, and seen by inflicted workers as sincere or suspect due to the physician's possible conflict of interest. Except in cases of obvious injury, medical confirmation does not completely dispel the ambiguity surrounding on restriction.

Conclusion

To summarize the position developed here: workplace injury and illness are a major problem facing TMM and a significant aspect of the reality faced by production employees of Toyota. Every interviewee associated with production was either disabled at some time, knew someone who was disabled, or had his or her work load and/or work-team morale affected by disabled work mates. All were involved in kaizens or other safety programs to prevent disabilities. Workplace disabilities place a severe strain on the proper functioning of work teams. In some circumstances, disabilities can precipitate a complete breakdown of work teams. The most problematic type of disability is carpal tunnel syndrome because it results from the "normal" performance of the job, has a gradual onset, is difficult to cure, and receives an ambiguous response from coworkers. Thus, workers who suspect that they have a repetitive motion illness like carpal tunnel must weigh whether or not they can tolerate the pain, the possibility of permanent damage resulting from continued aggravation, the reaction they will receive from coworkers, the increased burden their restriction will cause teammates, and the likelihood of getting transferred to a less injurious job before they seek the dubious sanctuary of going on restriction. The disincentives to going up front with an injury lead to underreporting, working with pain, self-imposed restriction, and strategies to perform the job while disabled.

Although Toyota benefits in some ways from the disinclination of workers to report injuries and illness, it must be perceived by workers as vigorously involved in protecting their health and safety even if it means tolerating some false reports and malingering. I maintain that acceptance of some malingering is not an irrational policy. Instead, through its positive affirmation of the company-team ideology, it actually serves as an effective deterrent to malingering and tends to keep disabled workers on the line as long as they can stand it. Disabilities pose another threat to company team if improved physical conditioning of employees and aggressive management preventative efforts do not reduce the rates of injuries and illnesses. Workers will either come to doubt the "aggressiveness" of management efforts or they will begin to question the "necessity" of the lean work force and/or the "inevitability" of the rapid pace of work. These last two are major components in the company-team ideology and the first is related to the "we are all in this together" tenet. Doubting these aspects of the ideology would be a severe threat to the Toyota team itself.

6

ON AUTOMATIC

No analysis of work in a factory would be complete without some discussion of the repetition and tediousness inherent in most factory jobs. Sixty-seven percent of TMM's employees are team members, and the majority of them work in production. All production team members interviewed mentioned the boring quality of their work. Tom Lemming commented:

> On the line, once you get set into that routine and that particular flow, the work is very monotonous. We had twelve processes in our area and I could do ten of them easily without thought. In other words, step in there and do my process, one right after another, one right after another, one right after another, and you find yourself contemplating problems at home: what if I'd done this, what if I do that, or what am I going to do next weekend, and so on. Your mind just travels in a circle. I just felt idle. Although I was working physically very hard, my mind was idle.

Another production team member expressed it this way:

> Sometimes you find yourself like a robot. You say "Did I do that on the last product?" And you walk up to it, and yup, perfect as can be. But it becomes such a habit that your mind is not involved and you can't remember if you did something or not.

I asked Bob Dottard if his job was hectic. He replied:

> Not once you get used to it. In fact, its kind of relaxing to me now. I guess it was a little hectic when I was first learning it.

I can do any of the twelve jobs in the area now and think about where I'm going to go on vacation. Because really the jobs are boring. It's the same product every time. You're doing the same thing over and over 450 times a day. As you're in your job, you can kind of let your mind wonder off.

I called this phenomenon "on automatic" because this is the term used by the team members themselves to describe the condition and because it is a situation akin to flying an airplane on automatic pilot—physical processes continue, routine functions are carried on, but there is no pilot at the wheel. Through the team members' words quoted above we can see that they do their jobs automatically without engaging their conscious mind. There is nothing unique about this, it probably occurs at all large-batch manufacturing facilities. In fact it is a common characteristic of human mental functioning in general. We learn certain physical processes like typing, riding a bike, or driving a car and, after we have practiced these activities sufficiently enough that we are proficient at them, we can do them automatically, without thinking.

There is a definite advantage in learning these processes so that they can be performed automatically. We no longer have to think about how to do the physical processes, and can think about other things while performing them. Thus, we can think about the content of what we are typing instead of which fingers to put on which keys. We can think about the scenery outside instead of how much pressure to place on the gas pedal. Of course, the cost of this is that we become so comfortable with the physical process that we may not be alert enough to respond to the unexpected. To return to the automobile analogy, we may be so busy looking at the scenery that we fail to see the car that pulls out in front of us. Another cost is that once a certain physical process is habituated, learned to the point where it becomes automatic, it becomes very resistant to change. If we learned to drive an automobile with a clutch, we find we must concentrate diligently to prevent our left foot from slipping into action in a car with an automatic transmission. As soon as we relax our diligence, we find ourselves "resting" our left foot on the brake, at least until we have habituated driving with an automatic transmission. My analysis reveals these same dimensions of costs and benefits exist at TMM in regard to on automatic. As a result there is almost a dialectical tension between the organization's features that encourage on automatic and those that discourage it.

Company Policies That Encourage On Automatic

The majority of jobs performed at the TMM fall under the aegis of what employees called "standardized work." The principle of standardized work is that every motion, tool, position, and material of a job is minutely specified in an exactly prescribed time sequence. Directions on how the job is to be performed are written down, accompanied by a written explanation of the reasons why certain key parts of the job have to be performed in exactly the manner specified. The company logic is that if workers understand why a job has to be performed in exactly a certain way, they will be less likely to introduce "creative deviations." The job can be changed. Indeed, TMM encourages team members to change their jobs to make them easier or safer or to decrease labor or material costs, but only after going through official procedures. The job specifications are posted beside each work station so that if someone unfamiliar with the job has to fill in on an emergency basis they can quickly see how the job should be performed. Posting job specifications also assists in training new employees to the job and serves as a constant reminder to the experienced worker of how the job is supposed to be done.

In a factory where the performance of four thousand employees (at the time of the research) has to be coordinated in such a way that the same product is turned out one each minute with an acceptable level of quality, workers cannot be allowed to each do his or her own thing. Standardized work is an essential element in the coordination process. But standardized work assists in organizational goal achievement in yet another way: The exercise of performing the same motions over and over again increases the proficiency with which those motions can be performed. The muscles involved become stronger, movement becomes smoother, and eye-hand coordination increases. Every production team member interviewed indicated that once they learned their job it was no longer difficult to do and could easily be done in the time allowed, and they found a rhythm that decreased the mistakes they were likely to make. Lemming discussed with me the reason for TMM's position on standardized work.

TMM was concerned that people were not conforming exactly to their jobs. Some were starting their process early or working on it when the line stopped. They're afraid you'll leave something out in that situation. They want you to be in a real steady pace where its almost automatic. Get this done, go here. Get this done, go here. When people break the steady pace, defects occur. It's amazing how simple it is and how bad it can be when you don't follow it. I've been doing the same job for eight months. It's hard to focus directly on what you're doing. Being on automatic sort of makes up for any lack of concentration you might have.

Production workers become more proficient the more automatic the job becomes. To fully appreciate the proficiency of an automated skill, refer back to the automobile analogy. If you doubt that your driving is on automatic, imagine teaching someone how to drive. In order to teach the skill you would have to verbalize directions about, for example, how much pressure to put on the gas pedal and how far to turn the steering wheel under varying circumstances. This exercise would force you to dip into your automatic driving skills, those skills that just come naturally to you now, and consciously think about them—a task that's difficult for many. Comparing your driving skill to the skill of the novice in training should convince you that your driving proficiency is greatly aided by the fact of your experience, the fact that you can drive without thinking about it.

The downside of on automatic for workers, like automobile drivers, is boredom that prompts them to escape mentally into some other activity—planning vacations, weekends, and the like. Workers who are thinking about everything but their job will be less alert to an unusual situation that may arise, a defect or safety hazard. They may become complacent and comfortable with dangerous equipment and processes that could result in injuries, defects, and/or damage to the equipment. Therefore, TMM wants workers to operate on automatic while being mentally engaged in their job. This is apparent in a comment made by Fetalucci:

They want you to think about your job, to concentrate on what you're doing. They do want that. But it's hard to con-

centrate. They don't want you to get away from what you're
doing. . . . I've heard so many talks on this, I know I can find
something here in my mind to explain it.

TMM wants both: alert workers will catch defects, preventing injuries
and damage to equipment that occur with complacency, and workers on
automatic will be so proficient that they will cause fewer defects, injuries, and damage to equipment.

Further complexity is added by the fact that, unlike inanimate
objects, human beings may actively respond to boredom and monotony
in ways detrimental to organizational goal achievement. Humans often
invent mental challenges in jobs devoid of them. Hamper (1991), a
former GM worker, recounts how he and his coworkers reacted to
assembly line work:

> The one thing that was impossible to escape was the monot-
> ony of our new jobs. Every minute, every hour, every truck,
> and every movement was a plodding replica of the one that
> had gone before. (41)

And,

> Entire shifts would sail by during which I hardly developed
> a tangible current of thought. (92–93)

According to Hamper, workers responded by smoking pot and/or drinking alcoholic beverages before work, after work, and during lunch.
They devised games, rivet-shooting contests, and practical jokes to pass
the time. Workers "doubled up"—a process whereby two workers learn
each other's jobs. Then one worker does both jobs while the other
worker sleeps, reads, or carries on some activity off premises like
attending college classes, planting crops, drinking in a bar, and so on.
All the while both workers are clocked in and receiving full pay.

Roy (1952) and later Burawoy (1979) describe ingenious systems
created by bored and hostile shop workers to restrict output. Frederick
Taylor developed scientific management to deal with the goldbricking
he believed to be pervasive throughout factories and the working
classes in general. Thus, there is a real threat that Toyota's team mem-

bers might react to on automatic by devising systems for avoiding standardized work, by-passing safety rules and procedures, sabotaging products, or deviating from work discipline rules. They may refuse to actively participate in organizational activities like quality circles and kaizens.

None of my informants indicated that team members used strategies that could be construed as contrary to organizational goal achievement or hostile to TMM. Some told me of team members with "bad attitudes," but even these workers were not actively destructive. Instead, a bad attitude is manifested in refusal to participate in cleanup operations, group discussions, company, group, or team social functions, or quality circles. The high ratio of team leaders to team members, the company ideology, and the monitoring and sanctioning capacity of the work team itself act to deter destructive strategies and expressions. Still, the possibility of destructive worker reactions to on automatic provides a powerful incentive for TMM to strike a balance—to have just enough on automatic, but not too much. Toyota's lean production system makes it particularly vulnerable to the attitudes of assembly workers, and this gives those workers an inordinate amount of power compared to the position of their peers in traditional Fordist assembly plants.

Company Efforts to Mitigate On Automatic

Attempts to balance the dual goals of having workers on automatic and also being mentally alert in their jobs are exemplified in TMM's rotation system. To encourage on automatic, a team of workers is placed on a set of jobs and left there until someone is promoted, transferred, or terminates. Team members are not moved to a new job every day. In this way team members become proficient on the jobs they perform. To decrease on automatic, team members rotate within their team, or sometimes within their group, so that a different job is performed in each two-hour segment of the day, but the same four or twelve jobs are performed over and over again. This rotation system will require more time for workers to become proficient on all of the jobs they perform, but the addition to alertness and morale may more than compensate for the amount of time necessary for proficiency. Obviously, interruptions in daily work routine like lunch and coffee breaks reduce the negative affects of on automatic. I hesitate to cite these since, even though their impact is significant, TMM is required to provide these breaks and, therefore, cannot take credit for their contribution.

Other ways TMM hopes to keep workers alert while performing the same job over and over again are daily safety programs, emphasizing that kaizens and defect detection are integral parts of each production job, and promotion possibilities. I will make brief comments about each of these. The safety programs are intended as an antidote for the complacency that seems to naturally accompany proximity and repetition. Some groups have safety meetings twice during a shift and frequently the format is that team members are expected to present short talks or in other ways to be actively involved. Regular meetings at the group or departmental level also usually contain safety components. Team members participate in regular safety poster and slogan contests with money as a motivator for participation. It is hoped that the constant reminders of hazards and the importance of safety will make team members more alert to these kinds of problems and perhaps make safety an automatic response.

Programs intended to involve team members in improving their jobs and in upgrading the quality of the product—kaizens, quality circles, and constant quality monitoring—are, among other things, efforts to broaden the job beyond the physical tasks prescribed in the standardized description. If TMM is successful in motivating team members to participate in these activities, the team members will experience mental challenges during the day. This may improve their morale and their overall feeling of importance to the organization. As a result they may try harder to stay alert at their jobs. Nevertheless, it will not change the tedium of performing the same tasks 450 times a day, or four sets of tasks each requiring 112 repetitions a day. Furthermore, if the company is so successful in motivating workers that, like one informant, they spend their time on automatic thinking up safety kaizens and redesigning jobs, are they any more alert to hazards and defects than if they were thinking about where they spent their last nickel? Job enlargement efforts like participation in kaizens and quality circles make on automatic more palatable to team members, and certainly contribute to organizational goal achievement, but do not change the numbing quality of assembly work.

Another organizational feature that increases tolerance for on automatic for some team members is the possibility of promotion. Since

TMM's policy is to promote from within, and since there have been three major expansions since start-up, team members feel there is a good chance that their total career with the company will not have to be spent working on the line. Time served as team member is seen as an investment by ambitious team members wanting to be promoted. The fact that TMM has been able to engender this hopefulness among some team members contributes to their willingness to work as proficiently as possible and, at the same time, be watchful for quality problems or unforeseen hazards. Putting up with on automatic is a type of deferred gratification for team members motivated by a desire for upward mobility. Thus, promotion possibilities may be an especially effective mechanism for countering the negative effects of monotonous work with the ambitious team members, who are most likely to be dissatisfied.

Informants discussed coping tactics they use to pass the time. Woulton enjoyed the solitude of her job and the fact that it doesn't "mess with my mind." She said,

> One thing I do like about the job they gave me in assembly is that I'm at the end of a line. You don't have anybody who bothers you during the day. You can do your job and as long as you do it well, everybody leaves you alone. I enjoy that because it gives me the opportunity to think of other things.

This response was unusual. A more common coping technique was socializing with coworkers:

> One good thing about our processes as a whole is that we can communicate throughout. When I'm building my gizmo, a guy comes to take the gizmo to the whatsit with me. We talk. When we are back there by the cupboards, we use the back doors, so there is someone there to talk to. It's more or less to break boredom. You've got to do something.

Implications

As we look at the implications of on automatic we are confronted with a number of questions. Will the balance that TMM has attempted to create continue to work without expansion and the lure of promotion possibilities? One Japanese executive noted that this was one of his primary concerns for the future of TMM. He believes that promotion possibilities are an important motivator at all hierarchical levels, not just for team members. Without continued expansion, promotion opportunities will be substantially reduced. And, he reasons, expansion cannot continue indefinitely. What then?

If TMM cannot offer viable opportunities for promotion for line workers, should the company continue hiring the most skilled (in terms of company measures of communication skills, manual dexterity, problem-solving abilities, and ability to work in a group), the most motivated and perseverent of candidates available in the labor pool? How long can such a capable work force be content working on the line? On the other hand, can the TMM production system involving cross training, kaizens, team work, flexible job definitions, and jidoka function without them? As an American manager at TMM sees it, this is the inherent paradox posed by the Japanese production system: the need for motivated, skilled workers for jobs that offer few intrinsic rewards.

In the process of transplanting the Toyota culture to the Kentucky plant, TMM may have brought with it some of the same contradictions that Japanese managed firms experience in Japan. Research consistently finds that Japanese workers express more dissatisfaction with their work and similar or less commitment to their employer than American workers, while at the same time working more steadfastly and diligently (Cole 1979; Lincoln and Kalleberg 1990). Simply put, Japanese workers have lower rates of absenteeism, turnover, tardiness, and militancy than do American workers even though the Japanese workers say they are less happy with their work. (This position is elaborated elsewhere; see Besser 1993.) The researchers who discover these results find them so disconcerting that they attempt to discredit their own data. They think it's impossible to have relatively dissatisfied workers (as compared to similar American workers) who work hard, check for quality defects, show up on time, are hardly ever absent, and who don't even take all their vacation time. Therefore, researchers deduce that there must be an error in their measurement of dissatisfaction and commitment attitudes.

In light of the Japanese system of organizing as exemplified in Team Toyota, the disparity between workers' attitudes about work and their actual performance on the job becomes plausible and understandable. TMM employees make a high-quality product. The Camry consistently heads its class of autos in quality and reliability (*Consumer Reports*, April 1995; J. D. Powers Initial Quality Survey rated the Camry number one in 1990 and 1991, *Car and Driver Magazine* ranked it the number one family sedan in 1993). In 1990 the Georgetown plant was awarded the J. D. Powers Gold Plant Award in recognition of Camry's ranking as the best car built in North America.

Absenteeism at Japanese auto plants in North America is even lower than at Japanese auto plants in Japan (5.0 in Japan compared to 4.8 in the United States according to Womack, Jones, and Roos 1990, 92). Turnover at TMM during the first couple of years after start-up was only 5 percent per year (Prather 1989). Elsewhere I recounted that workers run to help get the line back up and functioning and come to work when they are injured. They come in early without pay. They take job-related classes after work hours without pay. This certainly sounds like the behavior of a committed work force.

Yet we know that absenteeism, tardiness, and shoddy quality work on the part of a TMM employee will directly and immediately impact that employee's work team. These aren't just coworkers, but friends whose welfare and opinions matter. Therefore, does showing up on time, every day, working hard and looking for quality defects indicate commitment to the organization, or a need for group approval coupled with a fear of informal group reprisal? I think we can see that, in the case of TMM at least, employees could have relatively low attitudes of commitment to the organization and still perform as if they had high commitment. Additionally, consider the influence of the supervisory ratio—one team leader for every four or five employees—on committed behavior. There's no way Hamper and his fellow GM workers could get away with their high jinks at Toyota. Proponents of Japanese management would also contend that the despair and alienation personified in Hamper's behavior would also be far less likely to occur at Toyota.

This discussion is not meant to imply that TMM employees are not committed to Toyota and their work. Almost all the Toyota employees I have encountered appear to be very dedicated. The point I raise is that attitudes of commitment to the organization and feelings of job satisfaction, except in the extreme, may be irrelevant to job performance at a Japanese-managed manufacturing facility. If true, top-quality cars can be efficiently produced by workers who are less than enthusiastic about their job, so long as they are not openly hostile.

One other issue should be addressed before we leave this subject. The willingness of the Japanese to work as if they were committed even while expressing dissatisfaction compared to American workers occurs in a context of greater worker dependence on employers than is the case in the United States. The lifetime employment offered by the wealthy, prestigious employers is a double-edged sword. It means economic security until retirement, familiar surroundings (even if an employee relocates, the organizational culture, rules, hierarchy and personnel remain the same) and lifelong friendships with the same set of people. On the other hand, it also means that a job offer from a lifetime employer comes to each person only once. There are no midcareer changes to comparable or better Japanese employers, no searching for a better opportunity with another organization, no way to improve your situation by sending your resumé elsewhere. Once an employee accepts a "lifetime commitment" with a core organization, he (in Japan usually only men are offered lifetime employment) will be viewed as disloyal by other employers if he considers breaking his commitment by looking for employment elsewhere (Rohlen 1974; Marsh and Mannari 1976; Alston 1986). Only a lower-level organization would consider hiring such a "dishonorable" person. Indeed, even family and friends are likely to share this sentiment toward the employee. Workers do change jobs in Japan and they do so about as often as the Germans (Plath 1983); however, most of the movement is among the 70 percent of employees who work for the noncore, smaller, less prestigious firms, and among the temporary employees of the large firms.

Given these circumstances, imagine working in a Toyota assembly plant in Japan. Your dependence on Toyota, reinforced by family and friends and the lack of other comparable opportunities, would almost certainly influence your willingness to work hard even as you complain about Toyota. Kamata (1982), a Japanese reporter who worked for Toyota in Japan as a temporary assembly worker, documents just this kind of grumbling acceptance existing among many of his Toyota coworkers. TMM's workers in Georgetown are in a comparable, although less severe situation. There is no social stigma associated with changing employers in the United States, and friends and family are more likely to support a decision to quit a job. Nevertheless,

good-paying, secure jobs with decent benefits are hard to come by in
central Kentucky, even for college graduates. Employee dependence
may increase the possibility that TMM employees, like their Japanese
counterparts, will endure jobs with few intrinsic rewards, jobs that lead
to on automatic, and not resort to the diversions and hostility described
by Hamper, Roy, and Buraway.

Conclusion

In this chapter I have elaborated the condition of performing repeti-
tious, tedious tasks. A natural consequence of performing such tasks is
boredom and the ability to perform the job without thinking, which I
have called "on automatic." Toyota desires team members to possess
the manual proficiency that accompanies on automatic and at the same
time to mentally concentrate on the job. It has been my contention that
these are contradictory goals and so TMM's policies can only strike a
compromise between physical proficiency and mental alertness in jobs
that lead to on automatic. Specific TMM policies that address this issue
are job enlargement programs, safety awareness programs and job rotation. Less directly, company-team ideology, if effectively implemented,
will motivate team members to try to accomplish both goals. Moreover,
work teams and the low team-member to team-leader ratio act to prevent behavioral reactions to on automatic that are contrary to organizational goal achievement. Team members cope with the boredom of their
jobs by thinking of other things while their bodies automatically do the
tasks; socializing with teammates; developing an instrumental attitude
toward the job; and some convince themselves that line jobs are an
investment they must make to be able to be promoted.

On restriction and on automatic pose significant challenges to
Team Toyota. The nature of these challenges revolves primarily around
finding ways to motivate line workers to participate in the tacit agreement when growth has slowed and the possibility of promotions dimin-
ishes and after they experience years, and even decades, of line work.

7

THE WOMEN OF TOYOTA

This chapter represents a deviation from the material discussed thus far in the book. The previous topics were raised by the interviewees themselves. *Team, on restriction,* and *on automatic* were identified in numerous ways as the dominant features of work at Toyota's Kentucky plant. *The women of Toyota,* on the other hand, is not a subject that jumped out of the data. It is not one of the categories that was impossible to ignore—in fact, the subject of women in Japanese transplant organizations has been consistently ignored by researchers and pundits alike. The lack of serious attention given to the topic is unfortunate for two reasons. First, an examination of how gender issues are handled at Japanese transplants provides us with a mechanism to gauge the significance of the "Japanese" in Japanese management. In other words, considering gender issues will help us evaluate if there is a core management style that can be separated from Japanese cultural values (specifically, their values regarding women in the work place), and transferred to other countries. Toyota and the other auto transplants are gambling that Japanese management will work as effectively anywhere in the world as it does in Japan, even as they make some accommodations for local cultures and customs. The question remains open however. Can the Japanese leave their attitudes toward women and work at home and adopt policies and practices in their U.S. transplants that are consistent with American law and custom?

The second reason for the importance of the topic of women and Toyota is that women comprise a sizable percentage (25 percent) of Toyota's work force. If their experience, expectations, and opportunities are different from their male coworkers, how can work teams and company teams function effectively? There is some reason to suspect that women might encounter a different world at Toyota than men

151

encounter. Only recently in Japan have women been admitted in limited numbers to the ranks of permanent employment in large organizations. Even in the United States, few women have been employed as equals to men in heavy manufacturing plants. This suggests that Toyota's Japanese and American managers alike might view women employees differently than they view male employees, creating disparate realities for the two groups. If disparate realities result in women employees feeling that they are ill treated, they may become hostile and/or demoralized. This in turn will seriously jeopardize the successful implementation of work team and company team. I have stressed throughout the book the fragile nature of the Japanese manufacturing system. It is highly dependent upon the commitment, skill, and good will of line employees. Thus, anything that threatens the dedication of a large percentage of employees also threatens the efficient production of quality Camrys. It is because of the significance of gender issues to the generalizability of Japanese management as well as the nurturance of effective work teams and the company team that I have included a section on women and Toyota.

The Situation of Working Women in Japan

Before I elaborate the position of women at TMM, it will be useful to understand how women fare in Japanese organizations in Japan. Japanese management as a style of organizing human activity is gender neutral. The sex of employees is irrelevant. There is no provision for the sexual division of labor or for treating male and female employees dissimilarly. Japanese society however, like our own, is not gender neutral. In the West we have the impression that the majority of Japanese women are housewives who are not in the labor market. While the role of housewife is accorded more respect in Japan, the majority of Japanese women (57.8 percent in 1987 [Saso 1990, 4]) work outside the home in addition to their role as housewife. Women constitute almost 40 percent of the Japanese labor market (38.7 percent in 1992 [Department of Economic and Social Information and Policy Analysis, Statistical Division. 1994, 298]). Nonetheless, their labor market experience is quite different from their male counterparts and from female laborers in the United States. Women in Japan are far more likely to be self-employed—12.5 percent of the women in the labor market, compared to 5.7 percent in the United States—and 20 percent of women in the Japanese labor market work in family enterprises, compared to 0.8 percent of women workers in the United States (Brinton 1989). Also women in Japan are more likely to leave the labor force when they marry and especially when they have young children at home. They reenter later when their children need less attention.

Roughly equal numbers of women and men are hired each year in the large, core Japanese companies (Brinton 1989), which are the emphasis of this book. Technically, all these new employees—regardless of sex—unless hired explicitly as temporary employees, are offered the opportunity of permanent employment. In reality, however, there is so much social pressure exhorting women not to exercise their option to remain that, until recently, almost none stayed past their thirtieth birthday, or six to eight years experience with the company. The situation today is little better (Lam 1992). Brinton (1989) estimates that 71 percent of men who start with a core organization are on the career tract, compared to 23 percent of the women. Furthermore, the career tract for men is likely to be managerial, for women, clerical. In a study of forty Japanese firms in finance, insurance, and banking, 99 percent of the men were in the managerial career tract compared to 1.3 percent of the women (Lam 1992, 214–215).

There are several case studies that give us insight into the dynamics of women's employment in core Japanese organizations (Rohlen 1974; McLendon 1983). McLendon describes a Japanese trading company where half the employees are women, very few of whom are in supervisory positions. The large majority of women are hired straight from junior college. Since almost any high school graduate can perform the clerical and public relations tasks assigned to women employees, employers can select new employees on the basis of personal factors not germane to job performance. As a result, the primary selection criteria for female employees are their looks and personality. McLendon maintains that women's attractiveness is important for two reasons. First, attractive women increase the company's prestige with the general public and especially with clients. McLendon recounts a story about a group of employees from the trading company who were vacationing at a mountain lodge. The innkeeper greeted them by saying "You are from a big company, aren't you? Your women are very beautiful. Big companies always have beautiful women" (164).

The second reason for hiring pretty women with pleasing personalities is that they are viewed as a morale booster for company males:
"If young and pretty, they can create an atmosphere of feminine vitality
and freshness that will inspire the men to give their best efforts to their
work" (McLendon 1983, 164). A company with a reputation for hiring
attractive young women will be in a better position to attract the most
capable male employees. In addition to creating a pleasant work environment, employing large numbers of attractive women increases the
number of pretty women in the pool of potential marriage partners for
young male employees. This is quite important in Japan as there is little
opportunity for males to meet potential marriage partners elsewhere.
Teenagers and college students engage in infrequent socializing with
the opposite sex and practically no dating. Therefore, almost all newly
hired male employees are single with little experience with women. The
fact that their employer hires a large number of young, eligible, attractive women as potential marriage partners is an alluring benefit.

Several TMM employees reported to me that Toyota in Japan hires
two thousand new "office ladies" each year as potential wives for their
new male hires. Toyota in Japan is located in a rural prefecture, the significance of which is that if Toyota did not import these women to the
rural area there would not be sufficient numbers of available women for
their new male employees. In the process of hiring potential wives for
their male employees, Toyota has the chance to screen them and teach
them the company ideology and culture. This is beneficial to Toyota
because, as a result, the permanent employees are more likely to have
wives who will be sympathetic toward their husbands' long hours and
diligence toward work and who will eliminate all family and private
sphere "distractions" so that husbands can concentrate on Toyota business.

McLendon considers the issue of why women who have the promise of permanent employment with an important company would quit
after six or eight years. He explains that there are two possible scenarios. Either a woman marries, in which case she is expected to resign, or
she does not marry and is therefore, also expected to resign. The result
of both scenarios is the same, but the legitimating values, norms, and
sanctions are quite different. For married women, the rationale for res-
ignation is the belief that a person cannot give 100 percent commitment
to his or her employer (as employees of core organizations are supposed

to do) and at the same time give 100 percent to his or her family. In Japanese culture it is quite clear that, for women, the family must take precedence over any other demands. Therefore, a married woman's commitment to her family is suspect unless she relinquishes full-time employment. Under circumstances of financial necessity, she may work on a temporary or part-time basis, or in a family business that allows more flexibility. Nevertheless, Japanese values are unambiguous in communicating the inferiority of these choices. To lessen the chance of hiring women who might disagree with the prevailing views about the compatibility of home and work, employers can screen out women who might challenge the custom. Given the menial nature of the jobs women hold and the lack of promotion possibilities, it's little wonder that many prefer to be full-time housewives.

The second possible scenerio is that of a woman who is not married by age thirty. Even though her work commitment is not compromised by family demands, her attractiveness as a potential mate for male employees is greatly diminished. She is too old for the new hires and the males in her cohort are already married or searching among the younger women. In addition, she may be older than some of her male supervisors, making them feel awkward. As she continues to age without benefit of the seniority promotions enjoyed by the males of her cohort, the incongruency of her older age, which is supposed to confer respect and status in Japanese society, accompanied by her lower organizational status makes everyone uncomfortable. If she is not sensitive enough to realize the embarrassing position her continued employment represents to the company, her coworkers will remind her. She will be the subject of ridicule and ostracism. McLendon says that coworkers at the trading company called one woman who remained past her thirtieth birthday the Japanese equivalents of "aunt" said in a derogatory way, "old maid and spinster," and "widow who remains unmarried"—again, said sarcastically.

Women who remain after age thirty, whether married or not, may receive what is called a "katatataki" or "tap on the shoulder." This may be an actual or a symbolic organizational response to the woman who has overstayed her welcome with her employer. Supposedly the woman's supervisor will come up behind her while she is working diligently at her desk, tap her on the shoulder and say something like "isn't it time you resigned."

McLendon and Rohlen conducted their research in the late 1970s. More recent work (Lam 1992; Brinton 1989) indicates that the situation, at least statistically, has changed somewhat. Now 7 percent of all new female hires are considered to be on a career tract, compared to 22 percent of the males. Furthermore, more women are being hired in technical and professional positions than in the past. No doubt the labor shortage in Japan has motivated employers to take a second look at the vast labor potential of Japanese women and to adapt company employment policies accordingly.

Even if the history described by McLendon is slightly outdated, vestiges of it are, almost certainly, still present in Japanese organizations to greater or lesser degrees. Furthermore, this history gives us a glimpse of the world view, beliefs, and values of individual Japanese regarding the subject of women in the labor market. It may help to explain the frame of mind of Japanese managers as their organizations embark on establishing operations in foreign countries with foreign employees, laws, and customs.

Japanese Transplants

Considering the excitement generated by the transplantation of Japanese-style management to the United States and elsewhere, surprisingly little attention has been given to the subject of women in these organizations. One exception is the work of Mary Saso (1990), who studied women in Japanese organizations in the United Kingdom. Her research concentrated on employees at the NEC plant in Ballivor, Ireland, and the Matsushita facility in Cardiff, United Kingdom, but also discusses Nissan at Sunderland, U.K., and its Japanese suppliers located in the U.K. Saso concludes that there is no evidence that Japanese employers in the U.K. are treating female employees any worse than do other organizations. In fact, she reports that women working on the shop floor at NEC and Matsushita are quite satisfied with their employment and with their Japanese managers. She comments that Japanese managers have found female shopfloor workers in the U.K. to be unexpectedly diligent and hard working. For their part, the women's "image of Japanese managers is mostly of smiling faces whose owners are always fair" (191). Saso admits, however, that perhaps some of the reason for the high job satisfaction of women employees is that "they are just glad to have a reasonable job at all" (200). Both plants are located in areas with relatively depressed economies at the time of Saso's research.

What little information there is about the experience of women in Japanese auto transplants in the United States is often found in newspaper articles in reports of discrimination or harassment suits. For example, Diamond Star Motors (DSM), a Mitsubishi operation in Bloomington, Illinois, has been charged with sexual and racial harassment and discrimination by twenty-six present and former female employees. This case is unusual in that the claimants contend that the cultural values of the plant's Japanese management have contributed to the hostile environment toward women employees. The lawsuit maintains that during employee orientation, Japanese managers expressed the opinion that women do not belong in the work place and have created an atmosphere conducive to what the claimants' attorney calls the "the most gross and vile sexual harassment" (McKinney 1994). In its defense, Diamond Star maintains it has policies to ensure a comfortable work environment for all employees, including women and minorities, even though the DSM spokesperson was not sure that DSM even keeps records of the number of its employees who are women.

Honda U.S.A. does keep such records. Gelsanliter (1990) documents the early problems Honda had in meeting EEOC standards both in minority and female hiring. Currently 25 percent of the Honda U.S.A. work force is female and Susan Insley, who started as the vice president of corporate planning for Honda, is now the plant manager of Honda's engine plant in Anna, Ohio. Nissan U.S.A. has also been faced with charges of sexual discrimination. This pattern might suggest that sexual discrimination is so customary and "normal" in Japanese organizations that Japanese managers have been unable to separate the essential features of Japanese management from their own personal biases. This is precisely the position of the women claimants at Diamond Star Motors. Another possibility is that the treatment of women in Japanese transplants is no different from the situation of women in similar U.S. organizations. However, because employees expect a different kind of employment relationship with a Japanese company, women who encounter sexual harassment, even if it is no greater or more frequent than in U.S. organizations, may feel betrayed and express great anger.

Toyota promotes the corporate image that it is the best of the best. Accordingly, corporate managers are highly motivated to make TMM look better than its competitors and to avoid negative publicity. This is reflected in Toyota's willingness to use union labor in plant construction (although this decision was prompted by a few "embarrassing" union demonstrations), its initial hiring policies, sections in the first employee handbook (1988) condemning sexual harassment, financial contributions to the local school district, local universities, the town of Georgetown, human service organizations, and so on. Toyota had the advantage of arriving on the American scene after its Japanese competitors. It was able to observe the experience of Honda, Nissan, C. Itoh, and other Japanese firms that became embroiled in the embarrassment of civil rights litigation. It learned about American laws and practices from its own experience at NUMMI. Taken in this context, we would expect TMM's official policies and practices to conform to American laws regarding equal employment opportunity. At the same time, it's not surprising that these official policies are accompanied by awkwardness and clumsiness on the part of individual Japanese employees and instances of culture shock for everyone.

Women Employees at TMM

Women are found in all departments of TMM but are not evenly distributed throughout. Few women are employed in the stamping and maintenance departments. There are more in assembly, paint, and plastics, and women constitute the majority of staff in office and safety support positions. Although I do not have access to the actual numbers of women in supervisory positions, employees have told me about women who are team leaders, group leaders, assistant managers, and managers. In almost every issue of *Toyota Topics* one of the employees queried about vacation plans or recognized as a member of a team winning a quality slogan contest or appearing in the cameo spotlight that month is a female team leader, group leader, or specialist. At the end of this research in 1992, however, no women were above the rank of manager, which is five rungs below the president.

Toyota's official stance toward women employees is summarized in the *Team Member Handbook* (February 1988) in the section about sexual harassment. The section begins with the statement: "All team members should enjoy a work atmosphere free from all forms of discrimination, including sexual harassment" (94). Following this, sexual harassment is defined, TMM's policies toward it are outlined, and procedures for complaint are identified. It is worthy of note that policies and procedures about sexual harassment were in place before "Employee Assistance Programs" were implemented, before quality circles were organized, and before official policies about performance awards and sick leave.

Other organizational features reflect Toyota's official intention to smooth women's participation in Team Toyota. To facilitate the employment of women and men with parenting responsibilities, TMM offers multi-shift child-care services. In 1993, Toyota opened the newly constructed Family Center that provides child care and recreational facilities on the TMM grounds. Since TMM's child care costs the parents about the same as that available in the local economy, many choose not to use it. Still it is especially handy for parents of children in school (a bus takes them to school from the center and picks them up afterward) and for second-shift workers.

A common practice in Japan is for male workers to go from work to a bar where team bonding occurs over a long meal and many drinks. It is expected that all members of a work team will participate. These same workers may golf together on weekends and even take company vacations together. Thus, it is not unusual for male workers to have more regular and closer interactions with coworkers than with spouses. Female employees, except as companions in their early unmarried years, are excluded from this important career- and network-building activity. Even if women were allowed membership in these groups, the circumstances of their lives would make it difficult for them to participate. Involvement of this magnitude with work mates is premised on the assumption that someone else is taking care of the family, the household, the community, and one's day-to-day personal needs like shopping, laundry, banking, and such. Married women with primary family responsibilities would be especially disadvantaged.

From the vantage of women and men with family responsibilities, it's fortunate that this practice was not transferred to TMM. The team bonding encouraged and financially supported by Toyota often occurs on site, or involves spouses and/or families. Picnics, Christmas parties, trips to baseball games, birthday parties, and group pizza lunches are examples of company-sponsored, official, team-building events. Of course, individual employees might choose to go to a bar for happy hour after work, but it is not a common part of normal team-building efforts.

Let there be no doubt, however, TMM has had problems creating an equal opportunity employment environment. Still, I contend that the Japanese behavior at TMM toward women employees differs only in degree, not in kind, from what one would find with many American male managers in similar circumstances. We should anticipate that Toyota's Japanese managers and trainers, who have little or no experience working with women employed in nontraditional jobs, would be unable to perfectly implement Toyota's official position on their first try, even if they were personally dedicated to doing so. Therefore, regardless of official policies and intentions, the juxtapositioning of Japanese beliefs and a fairly significant number of women in nontraditional jobs has caused unusual experiences for both Japanese and Americans.

To illustrate: two managers who were among the first Americans hired described to me their disbelief when they examined the drawings for the plant layout. The facility included separate locker rooms for men and women complete with showers, toilets, and changing areas, so that employees could clean up after work. The problem was that the rooms were enclosed on the sides, but not at the top. Anyone using the catwalks above the locker rooms would have an unrestricted view of the whole area. It never occurred to the Japanese designers who planned the facility that there might be a need for privacy in the locker rooms. At Toyota plants in Japan, these areas are used exclusively by men. Also the Japanese have different norms regarding the acceptability of public bathing than Americans do. Once the Japanese executives were informed of the difficulty, the facility plans were hastily altered and ceilings were added.

Another design feature that reflects the taken-for-granted maleness of the work force relates to rest rooms. The architectural plans did call for women's rest rooms, but not nearly enough to accommodate use by almost one thousand women during the fifteen-minute breaks allowed production workers. Inadequate restroom facilities was a common complaint among women team members. This, too, has been addressed with the addition of more facilities. However, building too few women's rest rooms seems to be less a characteristic of Japanese culture and more a predilection of male architects and designers in general. To wit, American managers did not notice this problem when examining the plant layout sketches, and many public areas in the United States (football stadiums, coliseums, theaters, amusement parks, etc.) suffer from this same oversight.

In a similar matter, neither Japanese nor American managers anticipated that uniforms that fit male employees may not work so well for women. It's not that the managers didn't plan a "woman's uniform." There are women's sizes and even a khaki skirt (which for safety reasons can only be worn by office staff). It's just that many women find the pants to be uncomfortable and consequently don't wear them. If the purpose of the uniform is to lessen hierarchical differences, create a sense of equality among employees, and nurture the company team, ill-fitting pants for women becomes a more important issue than one of simple discomfort. Wearing different clothes could serve to set women team members apart from the rest of the employees. Women's lack of uniform clothing is less remarkable, though, given that many American male team members likewise wear variations of the company uniform. At the same time, the Japanese themselves are steadfast in their loyalty to the uniform. A possible consequence of Japanese managers consistently wearing uniforms was reported by a woman specialist:

> Since the Japanese are physically small men and always wear the uniform, they're not so intimidating. So you feel more confident in dealing with them. It's much easier to present the facts and answer questions when you don't get all hung up with "Oh my God. I'm talking to the president."

The interactions between Japanese male employees and American women in nontraditional roles reveals much about the sex-role expectations of both cultures. With qualifications to be discussed later, my female informants generally share the view expressed by women employees of NEC and Matsushita in the U.K.—that is, "Japanese managers are mostly smiling faces whose owners are very fair" (Saso 1990, 191). Unlike employees at the much smaller U.K. plants, however, TMM's production team members have had only infrequent contact with Japanese managers—except at the time of plant start-up. During that period, a large number of Japanese Toyota employees were temporarily assigned to the plant to train the new American employees. The

perception of Japanese employees held by many team members stems
from experiences occurring at that time. A production team member
described to me one such experience:

> Someone stopped the line. Our team leader (who is a
> woman) was standing around talking to another employee.
> A Japanese trainer came up to her and said, "You go fix.
> You go fix." She got mad and tried to explain that the prob-
> lem was not in her area and maintenance was tending to it.
> He started pushing her to get her moving toward the area of
> the problem. She turned around and pushed him back. He
> was so shocked, he just stood there. He couldn't believe that
> she would push him.

To illustrate a point, reread the description of the incident, but this
time take out the word *Japanese* and amend the broken English. Now
the experience is with "any man." Ask yourself, "How would an American male trainer have acted in this situation? Would he be less likely to
push the woman in the first place? It's fascinating that neither the teller
of this story, nor I, the listener, attended to this aspect of the trainer's
behavior, concentrating instead on his shocked reaction. My informant
didn't react as if she thought the trainer's pushing was unusual. Returning to the example, would an American trainer have been shocked to be
pushed back? Almost certainly. Would he have been shocked into inaction, as presumably the Japanese trainer was? Would he have been more
likely to retaliate in kind, or retaliate in a different way with a reprimand
of some sort? What effect, if any, does the likely larger size of the American male trainer have?

This mental exercise helps to separate the influence of cultural differences from the impact of sex-role expectations shared by both Japanese and American cultures. I maintain that the critical element in the
above story is the disjuncture of "any man" and assertive woman, and
not Japanese man and assertive American woman. In other words, culture seems to be less important in explaining the trainer's reaction than
the violation of gender roles generally accepted by both cultures. The
effect of culture is to change the degree of shock, but not the fact of it.
In the world view of the Japanese trainer, an assertive response by a
woman toward a supervisory male in a work place is not just unlikely,

as it would be for an American trainer, but it is unthinkable, unbelievable.

Indeed, American male managers were in for a little "culture shock" also. Production group leader, Ron Prichard, told me about a female team leader he hired:

> Coleen was my first team leader. When I first got her I thought, *Holy Cow, what have I done? I don't know beans about cars myself, and who's my first team leader, a waitress. What am I going to do?* The first day she showed up, I had just come back from Japan and I thought, *I've got to teach this girl something.* So I started to go through this very patient technique, hold it just like this, and then do that. . . . She said, "Give me that, let me try it." She whoops it in there—zip, zip zip—and she had it installed. Just like that. She's an extraordinary, outgoing, self-confident individual and she'll do well.

We wouldn't expect all female production workers or team leaders to be as capable as Coleen apparently is. What is startling to Prichard is that any of them are. Coleen is especially stereotype busting since she is a small woman who previously worked as a waitress. Given her background, would a Japanese group leader have reacted any differently? The attitudes displayed in this example seem well within the range of individual variation for either Japanese or American cultural expectations. Actually, Prichard indicated a great deal of openness in revealing his bias to me and seemed honestly pleased to have his doubts about Coleen proven wrong.

An important point highlighted by this example is that most TMM employees, except for the higher-ranking production managers, did not have manufacturing experience. Males and females alike were ignorant about manufacturing cars. Their former employers—that is, government, education, services, fast-food restaurants, and the like—were more likely to have a larger proportion of female employees—even if in lower-level positions—than traditional automobile manufacturing facilities. We would expect many of these men and women to have previously worked with members of the opposite sex as peers. Furthermore, men and women started at TMM together, went through assimilation together, some traveled to Japan together. Women were not interjected as an afterthought into groups of males bonded together by years of working side by side.

All of these factors not withstanding, some female employees maintain that sexual discrimination exists. A female employee working in a nontraditional capacity commented:

> I never considered myself a feminist until I started working for Toyota. Where I worked before, I never was made to feel as though they didn't trust me because I was a woman. But when I came to Toyota, a lot of us started new at the same time. I could see that men who came in at the same time I did were trusted much more quickly and were made to feel that they could handle responsibility much sooner than I was.

Another female interviewee, without prompting, reported an incident of sexual discrimination that she said was already in litigation. In casual conversations, other TMM employees mentioned rumors about two different situations, which, if true, would both meet the definition of sexual harassment. The veracity of these accounts is important, but even if untrue, reports of discrimination and rumors of harassment if repeated often enough can do damage to the TMM company team. If women employees—a quarter of the work force—doubt they'll get a fair deal, and feel more imperiled and less secure than their male coworkers, will they be willing to give 100 percent to the company?

In summary, the TMM work place is probably not free from sexual discrimination and harassment. Even so, it may be better and is almost certainly no worse than many other large manufacturing plants. I have walked through manufacturing plants where even visiting women on official tours are subjected to a barrage of harassing shouts and derogatory comments from line workers. This does not happen at TMM. Official policies may be less responsible for the friendlier atmosphere than the facts that TMM line workers are too busy to engage in such "entertainment," there is closer supervision than at most other manufacturing facilities, employees are younger and better educated, and there are more women working on the shop floor, some of whom are shop-floor supervisors. Whatever the reason, American women face a very different formal and informal organization at TMM than the picture I presented earlier of what Japanese women encounter in the work place in Japan. At least in this regard Toyota managers have been able to disengage Japanese management from Japanese culture.

The Wives of Japanese Expatriate Managers

Team Toyota, as it is actualized in Japan, would not be possible without women who are as committed to their role as wives as their husbands are to making quality cars. The practices of employees working long hours, working on Saturdays, and socializing with coworkers after hours are premised on the assumption that employees have a capable, dedicated wife to take care of everything else in their lives. As a rather extreme example, I present my Japanese friend who lays out clean clothes for her husband every morning, complete with the money he will need during the day, a handkerchief and car keys placed in the appropriate pockets, his pen and pencil holder in one shirt pocket and a fresh pack of cigarettes in the other. If she doesn't give him a coat or umbrella, he'll go off without them. An employee with a wife who sees to all of his personal needs in this way is able to focus all his energy and attention on the production of Camrys.

At TMM where there are both male and female permanent employees and where greater allowance is made for employee personal and family time, the impact of wives fulfilling this traditional support role is less that it is in Japan. Nevertheless, the Japanese wives present in Kentucky free Toyota's Japanese executives from concerns about the "private sphere" of their lives, assist in nurturing the company team by facilitating social interactions between Japanese and American employees of Toyota, and serve as informal cultural and corporate ambassadors to the community. These latter functions are not required of them in Japan.

For five years I regularly interacted with a number of Toyota Japanese wives. We attended the same Toyota functions. We lunched together, shopped together, spent holidays together, vacationed together, and visited back and forth in each other's homes. I belonged to an organization whose goal was to link American and Japanese women in the Lexington area. In this forum I met wives of non-Toyota Japanese managers. On a monthly basis this group discussed a wide variety of cultural patterns, including subjects like family, marital, religious, and parenting practices. My experience has left me with great respect and affection for Japanese women. Further, I became aware that while Japanese wives make substantial personal sacrifices and expend considerable effort in ways that are beneficial to Toyota, their contribution is almost totally unnoticed beyond personal, private conversations.

The role of Japanese wives is generally unrecognized and unexamined by analysts of Japanese management. In all fairness to the Japanese management scholars, it's common in studies of organizations to concentrate on internal features, and the contribution to organizational success made by wives of managers has usually been ignored by all organizational scholars (one exception is Kanter 1977). I too have thus far confined my attention to the internal features of Japanese management, ignoring subsidiaries, marketing, competitors, government support, and the like. These external elements are examined in other research and publications. However, as is the case of women employees in Japanese transplants, very little has been written anywhere about Japanese wives at transplants. It is important to at least introduce this topic and thereby increase the chances of future in-depth consideration. Therefore, in this section I will go outside the plant walls to present an overview of the situation of the wives of Toyota's Japanese managers.

Before I proceed, a few remarks about American cultural values are in order. The point of this detour will be clear shortly. In my Introductory Sociology class, I show a video on the Japanese educational system. The video prompts many comments from students about the rigor and focus of Japanese schools. Predictably, most think Japanese youngsters are required to work too hard and study too much. However, every semester students ask a question that astonishes me. They want to know, "Why do the Japanese bother to educate their girls so well when all they're going to be is housewives and mothers?" The students believe this practice is a waste of educational dollars and human potential. Now, we know that Japanese women do indeed work in the labor market, contrary to common misinformation about the subject. It is clear though that Japanese women's labor market participation rates are beside the point. The significance of the question is not that it illustrates ignorance about Japanese women workers but that it reveals American students' assumptions about the low level of skill and knowledge required to be a housewife and mother. Even more telling, it reveals the attitude that societal resources invested in giving mothers an elementary and high school education equal to that provided to the highest-ranking professionals are wasted.

The students' question lays bare our cultural devaluation of the role of housewife and mother. A view that, according to my observation and the conclusions of researchers who are specialists in this topic, is not shared by the Japanese. For example, Imamura (1987) cites several studies that, taken together, conclude that the Japanese hold the role of housewife in much higher regard than is the case in the West. One study argues that, in addition to the higher status accorded to "family work" in Japan, Japanese women are not convinced that men's lot in life is preferable to their own. In particular, they are skeptical that white-collar salarymen have a better life than they do. Joy Hendry (1993), who conducted three extensive participant observation studies of Japanese housewives (two in Japan, and one in the U.K.), similarly concludes that these women would be reluctant to change places with their husbands. They enjoy a freedom and flexibility in their lives that their husbands do not have. Hendry contends that Japanese housewives feel privileged to be in that position. She adds that neither they nor their families regard them as second-class citizens.

I believe many American observers of the status of Japanese women are influenced by the prevailing low opinion of homemaking in American society. Therefore, we think that because most Japanese women are expected to build their adult lives around their family role, even to the point of forced early "retirement" from paid employment, that they are coerced into an unrewarding, mundane, second-class existence. Put another way, American negative bias regarding the role of housewife, and nonwage labor in general, distorts our analysis of the status of Japanese women.

I was continuously confronted with my own prejudice on this matter as I grew to know the wives of Japanese managers in Kentucky. My first encounter with a Japanese wife was when I accompanied Mita as she shopped for furniture to fill her American house. She had been in the country for less than a week and knew little English. I assumed that she was "just looking" at furniture on this first trip out. A lot of money was involved and I was sure that she would need her husband's approval or, minimally, his knowledge, before making such a major expenditure. Much to my astonishment, before the day was over, she had purchased over $3,000 worth of furniture. I was afraid that she didn't understand what she was doing, given that she and the sales clerks knew only a few words in common. I asked her several times, "Don't you think you should talk this over with Sunitason (her husband) before you decide?" "No, no," was her reply each time, "Sunitason too busy." I learned later, that Japanese wives are typically the family financial manager. The husband gives his check to his wife, who in turn provides him with an allowance for his daily needs. The wife is responsible for all financial decisions, making sure that the household is suitably taken care of and that enough money is saved for major purchases, retirement, and children's education. It would be a mistake for Americans to make too much of this. While there certainly is a degree of power that accompanies this responsibility, it may be more a situation where household financial matters are seen as too trivial for husbands to be concerned with.

It goes without saying that Japanese wives encountered very different circumstances in Kentucky than what they were used to in Toyota City, Japan. Their feelings of disjunctive resulted partially from immersion in the American culture, but culture is not the only changed element they experienced. Toyota in Georgetown is not the same as the Toyota they were used to in Toyota City. Some of these women had been office ladies. They were married to Toyota salarymen. Some had lived a large part of their married life in Toyota housing surrounded by the families of other Toyota employees. Toyota permeated every part of their day-to-day lives. This new Toyota, the one adapted in "nonessential" ways to American values and norms, put more emphasis on couple and family events instead of same-sex socializing and bonding. At TMM, Japanese couples and families vacation together, an uncommon practice in Japan. TMM husbands and wives are expected to attend events together and entertain Toyota guests together in their home. The after-hours interaction between Japanese and American employees, stressed as an important element in building the company team, happens as often at couple events, or in a Japanese home, as it occurs on the golf course or a bar. In Japan, employees do not go to each other's homes and may never have met a coworker's spouse. Employees take business guests to restaurants or inns but not to their homes. Thus, for some Japanese wives, this was the first time they had been called upon to plan, prepare, and participate in business-related social events.

To better frame this couple activity, several of my Japanese acquaintances reported that their husbands continued to work while they, the wives, gave birth. One had not been out to dinner alone with her husband of twenty-five years since their wedding. Almost all were surprised to learn that American husbands purchase gifts for their wives for special occasions like anniversaries, Christmas, or birthdays. On the other hand, a few younger couples appeared to be quite experienced at socializing as a couple, no doubt reflecting "my familyism"—a current trend feared by some analysts to be undermining Japanese work ethics (Howard, Shudo, and Umeshima 1983).

Looking through the lens of American values about the importance of love and romance in marriage, we might think that Japanese wives would be happy with this new-found conjugal togetherness. In truth, it is difficult to know their feelings about this matter. I'm convinced that some wives would never divulge anything to an American that could be construed as critical of Japanese customs or Toyota. These women maintained that they were happy with their marital situation in Japan and happy with TMM's new expectations of them in the United States. For our purposes, their attitude is less important than the fact that they were required to make another adjustment at a time when it was necessary for them to make many.

I turn now to another role played by Japanese wives in the United States, that of unofficial cultural and corporate ambassador. Japanese assigned to overseas facilities have frequently been accused of establishing enclaves in those countries (Sethi, Namaki, and Swanson 1984; Tsurumi 1978). The enclave is used to provide both physical and social separation of Japanese families from indigenous cultures. Japanese in enclaves live close to each other, socialize almost exclusively with each other, and in some cases even establish separate Japanese schools for their children. This practice creates an impression among local citizens that the Japanese are aloof, quietly arrogant in their desire to barricade themselves from the pollution of other cultures. According to Toyota's Japanese executives, concern about Toyota's public image has prompted the company to discourage enclave formation. Japanese managers and their families are asked to scatter their households throughout the Georgetown and Lexington area, and they have done so. They enroll their children in neighborhood schools (although they did establish Saturday Japanese studies and language classes for children), and they join local civic and community groups.

Some Japanese wives, those with a good command of English, became actively involved in local community activities and groups. They teach Japanese language to Americans in informal classes, classes in Japanese cooking and Japanese music, and summer enrichment classes for American children. Japanese women orchestrated exhibits of Japanese culture at local museums and schools and became active in Parent Teacher Organizations, women's clubs, and international youth exchange programs. One local women's college had a pre-Toyota exchange program that brought in many female Japanese students each year. Toyota Japanese families acted as surrogate parents for the students. No doubt the support of the Toyota Japanese community in the local area improved the college's recruitment efforts. All of these activities, coupled with numerous others, help to create a positive image of Toyota among local citizens.

Many wives, however, were not fluent in English. These women could read English quite well, understand spoken English less well, and had a great deal of trouble speaking it. For this group, language difficulties and personal shyness precluded involvement in local social, community, or civic groups. They interacted almost solely with other Japanese, a very understandable response given their circumstances. They probably had no choice about moving to the United States. They had to give up friends and family, familiar neighborhoods, foods, and entertainment to spend years in the United States. They couldn't understand programs on television, movies, medical providers, their children's teachers, store clerks, or neighbors. Then there is the issue of what three to five years in the United States will mean to their children. Attendance at American schools will almost surely jeopardize the children's chances of scoring well on the critically important Japanese college entrance exams. Exposure to American culture and American peer groups during a child's impressionable years might make reassimilation into Japanese culture troublesome later on for the child. This is a very real fear frequently expressed by Japanese expatriates.

On top of all this, Toyota wants Japanese women to be sensitive to the image they project to the Kentucky community, and loosen their ties with each other enough to appear accessible and friendly to Americans. The fact that these women endured this situation with what, to an outside observer, appeared to be cheerful resolve is a considerable contribution both to the state of mind of their husbands and to Toyota's public image.

The personal accommodations made by Japanese wives and the contribution this represents to the successful implementation of Japanese management in overseas operations is an unexplored dimension of transplants. At TMM Japanese wives act as unofficial ambassadors of Toyota and Japan to the local community. They help to actualize company team by entertaining business guests and socializing with American Toyota families in a way that makes the Americans feel comfortable—that is, conforms to American rather than Japanese customs and practices. The impression among American employees that "we're all—Japanese and American—in this together" is likely to be enhanced by socializing in the homes of Japanese, eating food cooked by Japanese wives, listening to Japanese music, watching Japanese children and the interaction among and between Japanese families, admiring Japanese artifacts displayed around the house, and experiencing the attention of a Japanese hostess. Wives assist the realization of Team Toyota in all these ways in addition to the traditional role they play in facilitating their husbands' dedication to Toyota.

8

CONCLUSION

To conclude, I will return to the questions that prompted this research: How is Japanese management being transplanted in the American facilities of Japanese organizations and what does it look like to the American workers involved? Not all Japanese organizations have attempted to bring their style of management to their American facilities. However, Toyota is one of the few that is convinced that this style of organizing human activity is critical to their success and that it does not depend on Japanese workers for effective actualization. Using the amalgamated definition of Japanese management elaborated earlier, the extent to which Toyota has really transplanted Japanese management to TMM will be examined. In the process, how TMM's version of Japanese management is enacted at the level of TMM's employees will be summarized.

First, let us address the feature of Japanese management that is not present at TMM, the company union. Since company unions as they exist in Japan are quite different from unions in the United States, even if employees of TMM were unionized, it would probably not be in company union style. Toyota seems to have taken great pains to avoid working with American type union representation of its employees—for example, employee selection procedures, co-opting the union steward role within the organizational structure, great concern for the needs and desires of team members who are the employees most likely to unionize, and constant communication to team members about the fate of unionized employees in the auto industry whenever they are laid off or terminated. The result of these policies has been that, among my interviewees, no one was in favor of unionization and most were adamantly opposed to it. They did recognize, however, that some employees in certain departments were more receptive to the idea of unions due to their relatively worse work conditions.

One characteristic that all agree is a component of Japanese management is lifetime employment. There are no written, contractual guarantees of lifetime employment at TMM but neither is the arrangement guaranteed and written in Japan. Many workers at TMM assume they will have lifetime job security if they work diligently and do whatever is necessary to help Toyota achieve its goals. Employees expect that, in return for their commitment, TMM will be successful and, thus, their jobs will be assured. But more than that, employees expect that if circumstances lead to a downturn in the market, or if they have personal problems, Toyota will carry them through the bad times. The lean work force rationale buttresses belief in this tacit agreement, as does the culture and history of corporate Toyota. The true test of whether TMM honors the lifetime employment agreement will occur in the future when there is a downturn in sales or some similar circumstance that would normally prompt American firms to lay off workers. For the present, my informants believe they have lifetime employment and are willing to reciprocate by giving 100 percent in commitment, giving Toyota the benefit of the doubt and postponing immediate gratification for the promise of future rewards. American employees viewed the Toyota version of lifetime employment as obligating Toyota to stand by employees but not obligating employees to stay with Toyota should a better opportunity come along. This was especially true among the skilled, management, and professional ranks who are likely to have more opportunities to find comparable or better employment outside Toyota.

The job security promised by lifetime employment was very important to my interviewees. It was the reason given by employees who said they would choose to remain with TMM even if offered a better-paying job elsewhere. The lure of job security coupled with good wages and benefits, and the lack of other comparable opportunity locally, allows Toyota to hire the "cream of the crop" in the state labor market. Apparently this feature of Japanese management has much appeal to American workers, especially since they do not face the flip side of lifetime employment—the prohibition against changing employers.

Another distinguishing feature of Japanese management is the assignment of task responsibilities to groups. At TMM the work team is the actualization of group responsibility. All production and maintenance functions are assigned to teams. Most management and professional responsibilities are oriented around coordinating and/or supporting the work-team effort. We saw in the sections on work team and company team the importance Toyota attaches to group responsibility as evidenced by the extensive programs and structures intended to create and nurture teamwork. Further support for the importance of group responsibility is provided by the fact that special projects are carried out by teams—and by the assignment of semi-permanent teams—that cut across departmental and/or hierarchical levels, to quality control functions, to community and/or charitable responsibilities, and to the organization of employee social and recreational activities. Toyota seems committed to transplanting group responsibility to its American operations and, in so doing, validating group responsibility as a critical aspect of Japanese management. Indeed, it has been the contention of this research that team work is the primary explicating factor of work at TMM.

Job rotation is a characteristic of Japanese management that is present at TMM. It helps to build work teams, decreases the negative effects of monotonous work, offers a degree of job enlargement and, if done correctly, helps to prevent workplace disability. Interviewees talked about it in its capacity to prevent injury, and indeed, was seen by them as the primary solution to workplace disability.

Seniority pay and promotion systems are not present at TMM in the same form usually believed to be indicative of Japanese organizations. However, even in Japanese organizations in Japan, pay and promotion are based on complicated formulas with seniority having varying impact depending on the organization (Dore 1973). At TMM, pay is based on job category—that is, skilled team member, production team member, assistant staff, group leader, plant manager—more than on seniority. Within the category of team member, seniority makes a difference in pay up to the eighteen-month point for production and three-year point for skilled team members. After that time period, all pay is equal within these categories. In many respects this is a brazen departure from seniority-based pay to value-added-based pay. With eighteen months experience, a production team member is probably as competent as she or he will ever be at his or her job. In other words, according to this logic the twenty-year veteran of the production line adds no more value to the product than does the team member with eighteen months experience and, thus, should receive no more pay. Differential raises after eighteen months only come from increases in responsibility through promotion or transfer to another category like skilled team member. At the time of this writing, the most senior team members have less than ten years experience with Toyota. They usually came from jobs that paid less than their Toyota wage, and they know that Toyota pay is high compared to the prevailing local wage. All these factors support contentment with the pay system among team members. However, can that contentment be sustained as the workforce gains decades in seniority and as new employees are hired who after eighteen months make the same pay as twenty-five-year veterans? The seniority-based pay system present in Japan is supported by low entry wages that gradually increase as the workers' family responsibilities increase. If entry-level workers at TMM were paid comparably low wages to accommodate later increases for seniority, Toyota's ability to attract the "cream of the crop" in employees would be effected. We must ask also what impact a stratified pay scale for team members would have on the solidarity of the work team. Since the current pay system is intended to promote feelings of equality within the work team, what will be the consequences for work-team maintenance if seniority, or some formula to measure individual contribution, have a greater effect on pay? As long as TMM is expanding, the issue may not be a problem. The chances are, however, that at some point growth will level off and then TMM will have to confront the ramifications of their complete shift from seniority pay to pay based on the value of the labor to the product.

With other categories of workers, the pay system is more mysterious. There is a standardized pay schedule, but employees do not know where other employees started on the pay scale for their particular job category. As is frequently the case, these employees are secretive about discussing their pay with each other. Interviewees in these categories communicated to me that there was dissatisfaction among their ranks due to the low impact of merit on their pay. As a result, TMM has made accommodations to these groups allowing merit to be given greater weight in determining bonus pay.

Let us turn now from seniority's impact on pay to its impact in promotion opportunities. Promotion from the team-member to the team-leader category is based partially on seniority, partially on performance in the prepromotion classes and partially on the evaluations of group leaders and HR reps. Some team members are eager to be promoted and feel that merit should be even more important than it is in promotion considerations. The significance of seniority in promotion decisions at the management level is less well articulated in Toyota policy. It is suspected that seniority has less effect, especially since many American managers were hired at the same time.

The opportunity for upward mobility from the team member level to the management level is another area of interest. In many large Japanese organizations there are two separate career ladders: one accessible to high school graduates who start with the organization by working on the line and another ladder that reaches to the top of the organization and accessible only to those who are hired after their university education. This split career ladder is similar to that for noncommissioned officers and commissioned officers in the military. Some ambitious team members are unaware of the Japanese dual hierarchy and/or believe that TMM does not intend to import that feature of their Japanese operations to the United States. These team members citing the Toyota policy of promotion from within are looking forward expectantly to opportunities to move into management or professional categories. It is too early to know how open the higher positions at TMM will be to team members or, for that matter, how open the very highest positions will be to non-Japanese managers. The dual hierarchy that was a concern of several interviewees was the Japanese and American management hierarchies. If these dual structures continue to exist, neither seniority nor merit will be as critical in acquiring the top management positions as Japaneseness will be. Both promotion and pay have important implications for company team. Both must be perceived as equitable systems for the majority of employees to believe in the community of fate. Toyota will face significant challenges in the future as it attempts to negotiate between its corporate culture and the desires of its American work force. Currently TMM promotion and pay systems do not conform to the Japanese management model. Instead, Toyota appears to have adjusted the system to include consideration of merit, most obviously in the middle management, professional, and team member categories.

The benefits offered to TMM employees are a component of what is referred to in the Japanese management literature as "welfare corporatism." An organization practicing welfare corporatism looks out for employees in nonwork areas such as health, housing, and education programs for children. In return, the organization feels it has the right to be involved in personal employee matters like choice of spouse, behavior of children, and so on. The TMM benefit package is not as extensive as it is at their Japanese operations. For example, there are no dormitories or home loans for American employees here. Still, my interviewees believed the benefit programs were generous and reflected the fact that Toyota cares about employees, that it has a heart. In addition to offering a wide range of the usual benefits, TMM also provides PT money for social activities and has on-site child care and fitness centers. Contrary to what one might expect, given the assumption of American concern for individuality and self-reliance, most interviewees were actually very comforted by the belief that Toyota would take care of them.

As discussed in the section on the company team, workers frequently obscure the incongruity between the belief in self-reliance and the belief that Toyota will take care of them, by relying on the lean work force rationale and clauses of the tacit agreement. There was no indication from my sample that they felt that Toyota was intruding on their personal life, except insofar as work consumed a great deal of their time and energy and that some resented attending PT functions during off-work hours. There may be some subtle forms of intrusion of which the employees are not aware, however. For instance, as team leaders and group leaders socialize in PT functions with team members and/or become familiar with team members personally, they may be able to offer unobtrusive advice that is more effective in modifying personal behavior than commands from management. Undoubtedly, the potential for this kind of control of workers' personal lives exists in all organizations. But Toyota has designed a system—including low team-leader to team-member ratio, team leaders recruited from the team-member ranks and viewed as a lead worker instead of a manager, classes and policies to assure team-leader identification and commitment to Toyota ideology, and PT functions—which increases the potential subtle influence first-line supervisors can have on rank-and-file employees.

What we see at TMM is an abridged—and in some ways more elegant—version of welfare corporatism. Many of the usual benefits offered to Japanese employees are missing but also are the more obvious attempts to control employee nonwork behavior. The benefit package more closely resembles those offered by large American organizations than Japanese organizations in Japan. The distinctiveness from most large American organizations lies in the heightened potential for unobtrusive control of employees.

Consensus decision making is a feature of Japanese management that Toyota has brought to Georgetown. Interviewees did believe that, in comparison to the other places where they had worked, there was something different about Toyota in regard to how decisions were made. They reported that employees have more input into decisions at TMM than they had with their former employers. On the other hand, Prichard, a group leader, made it clear that even though consensus decision making is the ideal, in actuality most ideas in his group originated with him and decisions were made by him. Huntington, a top manager, describes consensus decision making as a process whereby input is sought from workers whenever they would be affected by a decision. To some interviewees it means that decision making is slower and that a great deal of time must be expended in meetings. Slider, a group leader, complained about all the meetings he is required to attend, wanting instead to be left alone to get his "real" work done. He said that TMM stands for "Too Many Meetings."

Consensus decision making at TMM does not mean that the organization is run democratically. There is a chain of command, with differential decision-making prerogatives associated with position in the hierarchy. Managers are supposed to seek the input of workers. Managers are supposed to circulate information about a matter to peers, and possibly to subordinates, before a decision is made. But managers still decide, and they have the power to overrule subordinates' wishes. What makes Toyota's decision-making system different from the bureaucratic system is the intricate network of linkages transferring information upward, downward, and laterally throughout the organization; the Toyota philosophy that encourages consensus decisions; and the belief by employees (engendered by the company team) that their opinions will be considered by the decision maker. Additionally, if the decision area pertains to personnel matters, staff from the human resources department (viewed as a powerful advocate of rank-and-file workers) are always involved. Slider likens the HR rep to the union shop steward in that s/he defends the rights of employees against arbitrary management decisions.

At TMM certain matters are outside the purview of deliberation by employees and no accommodation to employee consensus will be made. Thus, the line begins exactly on time. Absences and tardiness are not tolerated, except in emergencies. The work team is a potent deterrent to employee absence and tardiness as the team must carry the missing person's work load in addition to their own. Each production job is minutely prescribed and deviations from standardized work are unacceptable unless official approval is given for changes. These and other issues of work discipline are not open to debate. Toyota's corporate management has decided unilaterally that this is the way TMM will operate. So we see that consensus decision making as a characteristic of Japanese management applies to TMM only to the extent that the input of affected personnel is supposed to be sought and is supposed to be seriously considered before decisions are made, that within the parameters prescribed by upper management adjustment can be made for employee input, and that most interviewees believe that decisions are actually made according to the procedure just described.

Japanese organizations are characterized as having tall, fine-grained hierarchies with short spans of control. TMM does have short spans of control at the team-member level, with one team leader for every four or five team members. Even though the ideal of team leader as lead worker is not realized in every team, none of my interviewees expressed the sentiment that they felt closely supervised—if anything, the opposite was more common. How is it possible that people performing the highly regimented work of large-batch manufacturing, with one supervisor for every four or five workers, do not feel closely supervised? Several organizational features—especially company team, work team, and standardized work—together contrive to control and monitor team members' behavior in a way that is likely to be less offensive to workers.

For other categories of employees, company team is the only operative element to supplement the control and monitoring by immediate supervisors. At start-up TMM had what many interviewees referred to as a "flat hierarchy"—that is, few levels of management. There were only three levels between production team member and plant manager, counting the team leader as one level. Since then, at least two levels have been added. Some of my informants have posited a plausible explanation that management was initially understaffed so that new levels could be created as needed to reward aspiring managers and to respond to the increased work load generated by expansion. Whatever the logic, TMM still does not conform to the Japanese pattern of tall, fine-grained hierarchies. However, if Toyota intends to maintain lifetime employment for employees, it must find a way to reward loyal employees and sustain employee motivation over decades. In Japan, tall hierarchies serve this function. Therefore, even though the initial Toyota structure is relatively flat, I would expect the hierarchy to continue to differentiate until it conforms more closely to the Japanese model.

A central feature of Japanese management is its ability to convince its employees that the organization is not just a place to make a profit, but that it is a community of fate with cooperative relationships and commitments. Clearly Toyota has attempted to reproduce this philosophy at TMM. Through almost every Toyota structure and policy, from hiring policies and assimilation, to company uniforms and the leveling of management perks, we can see that the underlying goal is to create the reality of Team Toyota. And for good reason, for it is only through the company team and the community-of-fate philosophy that work teams can be constructively utilized for organizational goal achievement. It is only through company-team ideology that the diverse and often conflictual goals of various job categories and hierarchical levels can be mediated and orchestrated into cooperative action. For the company team to be effectively actualized, there must be an understanding regarding job security for all employee levels and some tangible signs of commonalty of fate as provided by Toyota welfare programs.

Through this detailed examination of which features of Japanese management are present at TMM, it becomes obvious that Toyota has tinkered with some aspects of Japanese management—for example, seniority pay and promotion, company unions, and tall, fine-grained hierarchy—but others have been inviolable. Accordingly, the core features of Japanese management as seen by the management of Toyota are group responsibility as exemplified in work teams and cross-hierarchical teams, job rotation, and the creation of organizational ideology and feelings of community, that is, the company team. The other elements that Toyota has transferred, uncorrupted, to its U.S. facility from Japan are associated with work structure and plant layout and include standardized work, fast set up and retooling, just-in-time manufacturing system, jidoka, and kaizens. According to Dave Ohiu, a top Japanese executive for Toyota, production technology is critically important to Toyota but will only function successfully in an atmosphere created by work team and company team, to use my terms for these concepts. This position is elaborated in the "lean production" literature discussed earlier in the book.

I would like to close this book with a few comments about the ideological schism in contemporary scholarship on Japanese management. This work will almost surely be criticized by the scholars of the exposé genre persuasion for failure to be outraged at the intellectual and physical exploitation of workers which they think Japanese management represents.

I share with almost all labor scholars a predilection for organizations that provide an enriching and satisfying experience for employees. Who among us did not applaud Volvo's experiment at Uddevalla with self-directed craft teams? However, assembling cars in the craft team fashion is a costly enterprise and few can afford a car produced this way. Even some who can, choose a cheaper Camry, Taurus, or Saturn. The choice today for auto manufacturers is not between craft teams and the Japanese lean production system. It is between Fordism and lean production.

Let's not glorify Fordism in an attempt to demonize Japanese management. If you look at Fordism honestly, it's represents an ugly picture for workers. Even if we wanted to return to Fordist factories as described by Hamper and Roy where pay and benefits are generous, work discipline is slack, and job security is high except for regular layoffs, we could not. The social contract of the 1950s, 60s, and 70s between labor and management is dead. With the flight of capital to cheap-labor countries and competition from Pacific Rim companies, many manufacturers are downsizing, outsourcing, cutting wages and benefits, moving, adopting aspects of the Japanese production system, or some combination of these strategies. In this climate, choices of competitive manufacturing styles are quite restricted and craft team assembly is not a realistic possibility.

The Japanese production system is laden with contradictions. On the one hand, work at a Japanese plant is fast paced, highly standardized, and routinized. On the other, it is characterized by job rotation and job enlargement including worker responsibility for quality and job redesign. Japanese lean production is more exploitative of employees (in the sense that more surplus labor is recruited from employees for organizational goal achievement) and at the same time critically dependent upon employee good will. Japanese plants create a tightly coupled, fragile system reliant on the skill and commitment of rank-and-file employees. Just-in-time, pull manufacturing systems operated with a lean work force allow no margin of error, no buffer for mistakes. A missing part, broken machine tool, or disgruntled employees staging a slowdown or work stoppage could immediately halt production throughout the plant.

This is both the strength and the weakness of the Japanese lean production. It is a strength in that heightened sensitivity reveals problems and requires solutions that might be ignored in systems with plenty of buffer, inventory, parts, and people. It is a weakness because rank-and-file employees gain power as a direct result of the critical role they play in production. The system collapses (or reverts to a Fordist mode of operation) with low-skilled, unmotivated, or angry workers.

It is important here to revisit for a moment the historical and cultural context within which Japanese management developed. Until recently, employees in Japan were quite dependent on their employing organization for services that in the U.S. and Western Europe are supplied by the state—for example, retirement pension and housing loans. Lifetime employment for employees of large organizations and government meant that employees who left their employer would not be able to secure another job with comparable wages, benefits, and prestige. Furthermore, employees of core organizations who quit work are stigmatized by family and friends. This is a strong incentive to remain with an employer even while dissatisfied. Close supervision and work teams help to ensure that even disgruntled employees will put in their time and do no harm. Company unions are unlikely to offer a voice for worker discontent. Therefore, Japanese laws and cultural beliefs hold in check the power of rank-and-file employees that is created by the Japanese production system. Employees are as dependent on the organization as the organization is on them. Oliver and Wilkinson called this characteristic of mutual dependence in Japanese organizations akin to the strategy of "mutually assured destruction."

Employees of Japanese organizations in the U.S. and Great Britain are not as dependent on their employers as their Japanese counterparts. The equation of mutually assured dependence derived in Japan must be adjusted for overseas operations. The majority of Japanese employers—for example, Japanese banks, electronics manufacturers, and trading companies do not even try to export Japanese management to places where the equation does not apply. Toyota believes it must transplant its production system to be an effective auto manufacturer. At the same time, it seems to recognize that in the U.S. environment, empowering rank-and-file employees and nurturing work teams is a strategy fraught with danger. No wonder it has labored to create a viable company team. No wonder it has been willing to irritate higher-ranking employees to appease the rank and file. No wonder it stringently screens new employees.

Japanese management represents a significant deviation from the traditional Western style of management. Nevertheless, it is not a diabolically sinister system for exploiting workers, nor a panacea for improved employee commitment and satisfaction. The truth lies somewhere in between.

Appendix A

Interview Questions

What was your background prior to joining Toyota?

Why did you decide to apply for a position with Toyota?

Describe the process you went through to become employed.

How long have you worked for Toyota and what is your current position and department?

Describe your usual work day.

How is working here different from the other places where you have worked?

How much contact do you have with your coworkers outside of work?

Has working for TMM caused you to make changes in other areas of your life (such as home life, leisure, community activities, etc.)? If so, in what way?

What changes would you suggest or do you foresee Toyota making in the future?

What advice would you give someone who was considering coming to work for Toyota?

Would you quit your job with Toyota to take a better paying job with another company?

Is there anything else you can tell me to help me understand what it is like to work in a Japanese organization?

Appendix B

Glossary of Terms
Originating with Informants or Official Publications

Andon Board   The board of green, yellow, and red lights displayed in prominent areas throughout the plant, intended to alert management and maintenance staff to problem areas on the line.

Andon Cord   The cord (or button) located at every work station that is supposed to be pulled whenever a problem occurs. Pulling the andon cord lights up the andon board. First the yellow light is lit indicating that the worker at the station needs help. If the problem cannot be fixed immediately, the cord is pulled again lighting the red light and stopping the production line.

Assistant Staff   Clerical staff.

Associate Staff   An office position one step above clerical, but still below specialist. Associate staff perform fewer clerical duties and are given more autonomy and responsibility than assistant staff.

Jidoka   An aspect of the production system that requires the line to cease running whenever production or quality problems appear and to remain shut down until the problem is fixed. The line can be shut down either by "smart machines" or team members pulling an andon cord when a problem is detected.

Just-in-Time Manufacturing (JIT)   A system that keeps inventory at a minimum by requiring suppliers, including in-house suppliers, to deliver daily, or even more often, only the supplies that are immediately needed.

Kaizen   A suggestion for improvement that will either reduce labor, cost, and/or waste or improve safety. This is part of the constant improvement system.

187

Kanban   A physical system of inventory control and production regulation. It is a tool utilized to realize just-in-time manufacturing. When the raw materials or parts that feed into the production and assembly processes are exhausted, a card indicating a need for these items is delivered to the preceding process. The preceding process then makes more parts, which are delivered to the succeeding process. When this batch of parts is used up, the cycle repeats itself. In this way no parts are made, no material is ordered until it is needed by the succeeding process in the production line.

On Restriction   A condition of reduced work load necessitated by a workplace disability.

PT   Literally, personal touch. The term is applied to company money made available to supervisors to use for employee social functions.

Specialist   TMM's term for professional staff in support positions, like accountants, engineers and computer specialists.

Takt Time   The amount of time between each finished product as they roll off the assembly line. A takt time of sixty seconds means one car is completed each sixty seconds. Also used to refer to the amount of time allowed to perform one set of tasks assigned each team member at a station. The series of tasks will be performed over and over again in a shift. The employee's takt time is related to speed of cars coming off the line, but may vary from it also.

References

Abegglen, James C. 1958. *The Japanese Factory: Aspects of Its Social Organization*. Glencoe, Ill.: Free Press.

Abo, Tetsuo. 1994. "Overall Evaluation and Prospects." In Tetsuo Abo, ed., *Hybrid Factory: The Japanese Production System in the United States*. New York: Oxford University Press, pp. 228–257.

Adler, Paul S. 1992. "The Learning Bureaucracy: New United Motor Manufacturing, Inc." *Research in Organizational Behavior* vol. 15: 111–194.

Advertising Supplement. 1988. *Lexington Herald-Leader* (Oct. 5).

Alston, Jon P. 1986. *The American Samurai: Blending American and Japanese Managerial Practices*. New York: Walter de Gruyter.

Armano, Matt M. 1979. "Organizational Changes of a Japanese Firm in America." *California Management Review* 21: 51–59.

Babson, Steve. 1995. "Whose Team? Lean Production at Mazda U.S.A." In Steve Babson, ed. *Lean Work*. Detroit, Mich.: Wayne State University Press.

Berggren, Christian. 1992. *Alternatives to Lean Production: Work Organization in the Swedish Auto Industry*. Ithaca, N.Y.: ILR Press.

Besser, Terry L. 1993. "The Commitment of Japanese Workers and U.S. Workers: A Reassessment of the Literature." *American Sociological Review* vol. 58 (Dec): 873–881.

Besser, Terry L. 1995. "Rewards and Organizational Goal Achievement: A Case Study of Toyota Motor Manufacturing in Kentucky." *Journal of Management Studies* 32:3 (May): 383–399.

Brinton, Mary C. 1989. "Gender Stratification in Contemporary Urban Japan." *American Sociological Review* vol. 54 (Aug.): 549–565.

Brinton, Mary C. 1993. *Women and the Economic Miracle*. Berkeley: University of California Press.

Burawoy, Michael. 1979. *Manufacturing Consent*. Chicago: University of Chicago Press.

Cole, Robert E. 1979. *Work, Mobility and Participation: A Comparative Study of American and Japanese Industry*. Los Angeles: University of California Press.

Cooley, Charles H. 1909. *Social Organizations: A Study of the Larger Mind*. New York: Charles Scribner.

Department of Economic and Social Information and Policy Analysis, Statistical Division. 1994. *Statistical Yearbook 1992*. New York: United Nations.

Dohse, Knuth, Ulrich Jurgens, and Thomas Malsch. 1985. "From 'Fordism' to 'Toyotism'? The Social Organization of the Labor Process in the Japanese Automobile Industry." *Politics and Society* vol. 14, no. 2: 115–146.

Dore, Ronald 1973. *British Factory, Japanese Factory: The Origins of National Diversity in Industrial Relations*. Berkeley: University of California Press.

Dreyfack, Raymond. 1982. *Making It in Management—The Japanese Way*. Rockville Center, N.Y.: Farnsworth.

Eisenhardt, Kathleen M. 1989. "Building Theories from Case Study Research." *Academy of Management Review* vol. 14, no. 4: 532–550.

Fucini Joseph, and Suzy Fucini. 1990. *Working for the Japanese: Inside Mazda's American Auto Plant*. New York: Free Press.

Garrahan, Philip, and Paul Stewart. 1992. *The Nissan Enigma: Flexibility at Work in a Local Economy*. London: Mansell.

Gelsanliter, David. 1990. *Jump Start: Japan Comes to the Heartland*. New York: Farrar, Straus, Giroux.

Gerth, H. H., and C. Wright Mills, ed. and trans. 1946. *From Max Weber: Essays in Sociology*. New York: Oxford University Press.

Gibney, Frank. 1982. *Miracle by Design: The Real Reasons Behind Japan's Economic Success*. New York: Time Books.

Glaser, Barney, and Anselm L. Strauss. 1967. *The Discovery of Grounded Theory: Strategies for Qualitative Research.* New York: Aldine De Gruyter.

Gordon, Andrew. 1985. *The Evolution of Labor Relations in Japan: Heavy Industry, 1853–1955.* Cambridge, Mass.: Harvard University Press.

Graham, Laurie. 1995. *On the Line at Subaru-Isuzu: The Japanese Model and the American Worker.* Ithaca, N.Y.: ILR Press.

Hamper, Ben. 1991. *Rivethead: Tales from the Assembly Line.* New York: Warner Books.

Hasegawa, Keitaro. 1986. *Japanese-Style Management: An Insider's Analysis.* Tokyo: Kodansha International.

Hendry, Joy. 1993. "The Role of the Professional Housewife." In *Japanese Women Working.* Janet Hunter, ed. London: Routledge, pp. 224–241.

Howard, Ann; Keitaro Shudo; and Miyo Umeshima. 1983. "Motivation and Values Among Japanese and American Managers." *Personnel Psychology* 36: 883–898.

Imamura, Anne E. 1987. *Urban Japanese Housewives: At Home and in the Community.* Honolulu: University of Hawaii Press.

Ishida, Hideo. 1981. "Human Resources Management in Overseas Japanese Firms." *Japanese Economic Studies* 9: 53–81.

Kamata, Satoshi. 1982. *Japan in the Passing Lane: An Insider's Account of Life in a Japanese Auto Factory.* New York: Pantheon Books.

Kanter, Rosabeth Moss. 1977. *Men and Women of the Corporation.* New York: Basic Books.

Kenney, Martin, and Richard Florida. 1993. *Beyond Mass Production: The Japanese System and Its Transfer to the U.S.* New York: Oxford University Press

Lam, Alice. 1992. "Equal Employment Opportunities for Japanese Women: Changing Company Practice." In *Japanese Women Working*, Janet Hunter, ed. New York: Routledge, pp. 197–223.

Lincoln, James R., and Arne L. Kalleberg. 1985. "Work Organization and Workforce Commitment: A Study of Plants and Employees in the U.S. and Japan." *American Sociological Review* 50: 738–760.

Lincoln, James R., and Arne L. Kalleberg. 1990. *Culture, Control, and Commitment: A Study of Work Organization and Work Attitudes in the United States and Japan.* New York: Cambridge University Press.

Lincoln, James R.; Jon Olson; and Mitsuyo Hanada. 1978. "Cultural Effect on Organizational Structure: The Case of Japanese Firms in the United States." *American Sociological Review* 43: 829–847.

Lu, David. 1987. *Inside Corporate Japan: The Art of Fumble Free Management.* Cambridge, Mass.: Productivity Press.

Mannari, Hiroshi, and Harumi Befu, eds. 1983. *The Challenge of Japan's Internationalization.* Japan: Kwansei Gakuin University and Dodansha.

Marsh, Robert M., and Hiroshi Mannari. 1976. *Modernization and the Japanese Factory.* Princeton, N.J.: Princeton University Press.

McKinney, Kathy. 1994. "26 Female Diamond-Star Workers Sue. Attorney Says Acts Were 'Vile Sexual Harassment'." *The Pantagraph* (Bloomington, Ill.) Dec. 16.

McLendon, James. 1983. "The Office: Way Station or Blind Alley?" In *Work and the Life Course in Japan.* David Plath, ed. Albany: State University of New York Press.

Milkman, Ruth. 1991. *Japan's California Factories: Labor Relations and Economic Globalization.* Los Angeles: Institute of Industrial Relations.

Monden, Yasuhiro. 1983. *Toyota Production System: Practical Approach to Production Management.* Norcross, Ga.: Institute of Industrial Engineers.

Nakane, Chie. 1970. *Japanese Society.* Berkeley: University of California Press.

Ohno, Taiichi. 1984. "How the Toyota Production System Was Created." Kazuo Sato and Yoshuo Hoshino, eds. *The Anatomy of Japanese Business.* Armonk, N.Y.: M. W. Sharpe.

Okumura, Hiroshi. 1984. "Interfirm Relations in an Enterprise Group: The Case of Mitsubishi." In Kazuo Sato and Yoshuo Hoshino, eds. *The Anatomy of Japanese Business*. Armonk, N.Y.: M. W. Sharpe.

Oliver, Nick, and Barry Wilkinson. 1992. *The Japanization of British Industry: New Developments in the 1990s*. Oxford, U.K.: Blackwell.

Omens, Alan E., Stephen R. Jenner, and James R. Beatty. 1987. "Intercultural Perceptions in the United States Subsidiaries of Japanese Companies." *International Journal of Intercultural Relations* 11: 249–264.

Ossola, Denis A. "Applications of Japanese Productivity and Quality Control Methods at Matsushita's Illinois Plant." In U. Krishna Shetty and Vernon M. Buehler, eds. *Quality and Productivity Inprovements: U. S. and Foreign Company Experiences*. 1983. Chicago Manufacturing Productivity Center.

Parker, Mike, and Jane Slaughter. 1988. *Choosing Sides: Unions and the Team Concept*. Boston: South End Press.

Ouchi, William. 1981. *Theory Z: How American Business Can Meet the Japanese Challenge*. Reading, Mass.: Addison-Wesley.

Pascale, Richard T. 1978. "Communication and Decision Making Across Cultures: Japanese and American Comparisons." *Administrative Science Quarterly* 23: 91–109.

Pascale, Richard T., and Anthony Athos. 1981. *The Art of Japanese Management*. New York: Simon and Schuster.

Perrrow, Charles. 1986. *Complex Organizations: A Critical Essay*. 3r ed. New York: Random House.

Plath, David, ed. 1983. *Work and the Lifecourse in Japan*. Albany: State University of New York Press.

Prather, Paul. 1989. Toyota Figures Show Most Workers Live Outside Scott County. *Lexington Herald Leader*, Nov. 15.

Reitsperger, Wolf. 1986. "British Employees: Responding to Japanese Management Philosophies." *Journal of Mangement Studies* 23: 563–586.

Rinehart, James; Chris Huxley; and David Robertson. 1995. "Team Concept at CAMI." In Steve Babson, ed. *Lean Work*. Detroit: Wayne State University Press, pp. 220–234.

Rohlen, Thomas P. 1974. *For Harmony and Strength: Japanese White-Collar Organizations in Anthropological Perspective.* Berkeley: University of California Press.

Roy, Donald. 1952. "Quota Restriction and Goldbricking in a Machine Shop." *American Journal of Sociology* 57: 427–442.

Saso, Mary. 1990. *Women in the Japanese Workplace.* London: Hilary Shipman.

Sethi, S. Prakash; Nobuaki Namiki; and Carl Swanson 1984. *The False Promise of the Japanese Miracle: Illusions and Realities of the Japanese Management System.* Boston: Pitman.

Shook, Robert L. 1988. *Honda: An American Success Story.* New York: Prentice Hall.

Strauss, Anselm L. 1987. *Qualitative Analysis for Social Scientists.* Cambridge: Cambridge University Press.

Sumner, William. 1940. *Folkways.* Boston: Ginn.

Terkle, Studs. 1974. *Working People Talk about What They Do All Day and How They Feel about What They Do.* New York: Pantheon.

Toyoda, Eiji. 1985. *Toyota—Fifty Years in Motion: An Autobiography by the Chairman.* Tokyo: Kodansha International.

Toyota Motor Manufacturing, U.S.A., Inc. 1st ed. Feb. 1988. *Team Member Handbook.*

Trevor, Malcolm. 1983. *Japan's Reluctant Multinationals: Japanese Management at Home and Abroad.* New York: St. Martin's.

Tsurumi, Yoshi. 1978. "The Best of Times and the Worst of Times: Japanese Management in America." *Columbia Journal of World Business* (Summer).

Womack, James P.; Daniel T. Jones; and Daniel Roos. 1990. *The Machine That Changed the World.* New York: Rawson.

Woronoff, Jon. 1983. *Japan's Wasted Workers.* 4th ed. Tokyo: Lotus.

Yoshino, Michael Y. 1968. *Japan's Managerial System: Tradition and Innovation*. Cambridge, Mass.: MIT Press.

Zelditch, Morris J., Jr. 1970. "Some Methodological Problems of Field Studies." In *Qualitative Methods: Firsthand Involvement with the Social World*. Chicago: Markham, pp. 217–231.

Index

A
advice cartels, 18
application process. *See* employee screening process
automatic, on, 140
    boredom and, 142–43
    efficiency increase and, 142
    promotion possibility and, 145–47
    rotation system and, 144
    safety programs and, 145
awards, performance, 92–94

B
bonus system, 92–94
bureaucracy
    baseball as an example of, 8–10
    Japanese management, comparison to, 16–17
    functionaries' role, 5
    principles of, 8
    weaknesses of, 10
bureaucratic flexibility, 101–4

C
carpal tunnel syndrome, 126–28
communication, open
    employees' attitudes toward, 98–99
    fostering belief in community of fate, 99–101
    role of hotline in, 97
community of fate
    creating in Japanese organizations, 14–15
    criticism of, 24
    nonmonetary awards role in, 94
    nurturing belief in, 88–89, 181
    role of open communications, 99–101

    values of individualism and, 105
company team, 51
    philosophy of, 86–87
    workplace disability and, 133–34
company union. *See* unions
conditioning, physical, 80–81
corporate team, 51–52, 111
    benefits to TMM, 112–13
craft team (Volvo)
    compared with lean production, 25
    as a costly enterprise, 182

D
decision making, consensus, 12–13, 179–80
disability, 120
    company-team ideology and, 133–34
    lean work force policy and, 134–35
    rate, 121, 123
    repetitive motion illness, 122–23

E
employee screening process, 56, 58–59
employability, enhanced, 116–17
employment agreement, 87–88
enterprise groups, 17

F
Fordism, 20, 182
frame of reference, prior, 78–79
frequency distributions, 28

G
general knowledge information, 30–31
goals, organizational, 67
grounded theory, 31–32

H
hierarchical distinctions
    leveling of, 90–92
    at TMM, 180–81
human resources, 94–97

I
individual importance, perception of,
    108–10
in-group out-group, 72
innovation-mediated production, 22
interviews, 33–34, 185

J
Japanese management
    decision making, 12
    division of labor, 13
    ideal type, 11, 16–17
    organizational culture, 14
Japanese organizations
    career ladders, 177
    success of, 17–18, 19–22
    women's employment in, 153–56
Japanese wives
    as homemakers, 167–68
    participation in enclaves, 169
    role of, 165–66
    as unofficial ambassadors at TMM,
        170–71
jidoka, 36
job rotation, 74–75, 144, 174–75
job security
    in a bureaucracy, 6–7
    Japanese model, 13
    lifetime employment, 149, 174
    role in community of fate, 104–5
just-in-time (JIT), 36

K
kaizens, 64-65

L
lean production system, 22–23

lean work-force rationale, 105–8, 182
lifetime employment, 13–14, 149, 174

M
maintenance team members, 42–45
management, Japanese. *See* Japanese
    management
management perks, absence of, 91–92
managers and specialists, 45–46
manufacturing plant, 38–39
methodology, research. *See* research
    methodology

O
on automatic. *See* automatic, on
on restriction. *See* restriction, on
open communication. *See* communica-
    tion, open
organizational goals, 67
    norms and sanctions effect on, 68–70
    perceptions of individual importance,
        110

P
pay systems, 175–76
performance awards, 92–94
personal touch (PT), 53–54
physical conditioning, 80–81
prior frame of reference, 78–79
prior manufacturing experience, lack of,
    59
production, lean. *See* lean production
production technology explanation, 20
promotions, 145–47, 176–77
pyramids, 11–12

R
repetitive motion illness, 122–23
research methodology, 27–28
    data sources, 32–34
    grounded theory, 31–32
restriction, on, 120–21
    carpal tunnel syndrome, 126–28
    disability prone jobs, 125–26

medical confirmation, 135–36
rate, 123–24
sanctions by work teams, 130
self-imposed, 130–32
stigma attached to, 128–29
up front, 127
rotation, job, 74–75, 144, 174-75

S
safety programs, 145
sampling, theoretical, 32
saturation, theoretical, 29
skilled team members. *See* maintenance
team members
slices of data, 32
standardized work, 141

T
tacit agreement, 87
and lean work-force rationale, 107–8
role of bureaucratic flexibility in, 103
role of performance awards in, 93
team concept, 2–3, 49–52
team leaders, 37
assimilation of, 62–63, 90
contributions to goal achievement, 71–72
problems with misunderstanding role, 82
reference group shift, 71
responsibilities, 54–55
team members
assimilation of, 60–62
attitudes towards work pace, 78
daily routine, 39–42
educational characteristics of, 57
motivations, 57
prior manufacturing experience, lack of, 59
problems with cooperation between, 82

problems with turnover, 83–84
work freedom, 66
team vs. individual, 76–77
theoretical sampling, 32
theoretical saturation, 29
Toyotism, 20
training programs, 44–45

U
uniforms, 39, 90, 161
unions
differences between Japanese and U.S., 15–16
reasons not at TMM, 173
role of human resources in preventing, 96
up front, 127

W
welfare corporatism, 15, 19
benefits as a component of, 177–79
women's employment
cultural expectations, 163
equal opportunities, 160–61
facilitation of at TMM, 158–59
in Japanese organizations, 153–56
sex-role expectations, 161–62
sexual discrimination/harassment, 157, 164
in transplant organizations, 156–57
work, standardized, 141
work agreement, 87
work improvement schemes, 75
work pace, changes in, 80–81
workplace disability. *See* disability
work team, 51
norms and sanctions of, 68–70
as primary group, 68, 174–75
problems with turnover, 83–84
relationships between members, 53